CW00508165

PLANNING AND CONTROL

USING

MICROSOFT® PROJECT 2013 and 2016

BY

PAUL EASTWOOD HARRIS

Windows XP, Microsoft® Project 2000, Microsoft® Project Standard 2002, Microsoft® Project Professional 2002, Microsoft® Project Standard 2003, Microsoft® Project Professional 2003, Microsoft® Office Project 2007, Microsoft® Project 2010, Microsoft® Project 2013, Microsoft® Project 2016, PowerPoint, Word, Visio and Excel are registered trademarks of Microsoft Corporation.

Primavera Project Planner®, P3®, Primavera P6 Project Manager®, SureTrak Project Manager® and SureTrak® are registered trademarks of Oracle Corporation.

Adobe® and Acrobat® are registered trademarks of Adobe Systems Incorporated.

Asta Powerproject® is a registered trademark of Asta Developments plc.

All other company or product names may be trademarks of their respective owners.

Screen captures were reprinted with authorization from Microsoft Corporation.

This publication was created by Eastwood Harris Pty Ltd and is not a product of Microsoft Corporation.

DISCLAIMER

The information contained in this book is, to the best of the author's knowledge, true and correct. The author has made every effort to ensure accuracy of this publication, but cannot be held responsible for any loss or damage arising from any information in this book.

AUTHOR AND PUBLISHER

Paul E Harris
Eastwood Harris Pty Ltd
PO Box 4032
Doncaster Heights 3109
Victoria
Australia
harrispe@eh.com.au
http://www.eh.com.au
Tel: +61 (0)4 1118 7701

Please send any comments on this publication to the author.

978-1-925185-30-0 – B5 Paperback
978-1-925185-31-7 – A4 Spiral
978-1-925185-32-4 – eBook

23 June 2016

SUMMARY

The book was written so it may be used as:

➢ A training manual for a two-day training course, or

➢ A self-teach book, or

➢ A reference manual.

The book has been written to be used as the basis of a two-day training course and includes exercises for the students to complete at the end of each chapter. Unlike many training publications, this course book may then be used by the students as a reference book.

This publication is ideal for people who would like to quickly gain an understanding of how the software operates and how the software differs from Oracle Primavera Project Manager, P3, SureTrak and Asta Powerproject thus making it ideal for people who wish to convert from these products.

CUSTOMIZATION FOR TRAINING COURSES

Training organizations or companies that wish to conduct their own training may have the book tailored to suit their requirements. This may be achieved by removing, reordering or adding content to the book and by writing their own exercises. This book is available in both A4 spiral bound, which lies flat on the desk for training and/or self–teaching, and in B5 paperback as a reference manual. Please contact the author to discuss this service.

AUTHOR'S COMMENT

As a professional project planner and scheduler I have used a number of planning and scheduling software packages for the management of a range of project types and sizes.

The first books I published were user guides/training manuals for Primavera SureTrak and P3 users. These were well received by professional project managers and schedulers, so I decided to turn my attention to Microsoft Project 2000, 2002, 2003, 2007, 2010, 2013 and now Microsoft Office Project 2016. This book follows the same proven layout of my previous books. I trust this book will assist you in understanding how to use Microsoft Project on your projects. Please contact me if you have any comments on this book.

SPECIAL THANKS

I would like to thank Martin Vaughn for reviewing my draft book and making valuable comments that I have used to improve the quality of this publication.

CURRENT BOOKS PUBLISHED BY EASTWOOD HARRIS

Planning Using Primavera Project Planner P3 Version 3.1 - Revised 2006
Planning Using Primavera SureTrak Project Manager Version 3.0 - Revised 2006
Project Planning and Scheduling Using Primavera Contractor Version 6.1 - Including Versions 4.1, 5.0 and 6.1
Planning and Control Using Microsoft Project 2010 and *PMBOK® Guide* Fourth Edition
Project Planning & Control Using Primavera P6 Version 7 - For all industries including Versions 4 to 7 Updated 2012
Planning and Scheduling Using Microsoft Project 2010 - Updated 2013 Including Revised Workshops
Planning and Control Using Microsoft Project 2013
Planning and Control Using Microsoft Project 2013 & *PMBOK® Guide* Fifth Edition
99 Tricks and Traps for Microsoft Project 2013 & 2016
Oracle Primavera P6 EPPM Web Administrators Guide
Planning and Control Using Oracle Primavera P6 Versions 8.1 to 15.1 PPM Professional
Planificación y Control Usando Oracle Primavera P6 Versiones 8.1 a 15.1 PPM Profesional
规划和控制 Oracle Primavera P6 应用 版本 8.1-15.1 PPM 专业版
Planning and Control Using Oracle Primavera P6 - Versions 8.2 to 15.2 EPPM Web

SUPERSEDED BOOKS BY THE AUTHOR

Planning and Scheduling Using Microsoft® Project 2000
Planning and Scheduling Using Microsoft® Project 2002
Planning and Scheduling Using Microsoft® Project 2003
Planning and Scheduling Using Microsoft® Office Project 2007
PRINCE2™ Planning and Control Using Microsoft® Project
99 Tricks and Traps for Microsoft Office Project - Including Microsoft Project 2000 to 2007
99 Tricks and Traps for Microsoft Project 2013
Planning and Control Using Microsoft® Project and *PMBOK® Guide* Third Edition
Planning and Control Using Microsoft Project and *PMBOK® Guide* Fourth Edition
Planning and Scheduling Using Microsoft Office Project 2007 - Including Microsoft Project 2000 to 2003 - Revised 2009
Project Planning and Scheduling Using Primavera Enterprise® – Team Play Version 3.5
Project Planning and Scheduling Using Primavera Enterprise® – P3e & P3e/c Version 3.5
Project Planning and Scheduling Using Primavera® Version 4.1 for IT Project
Project Planning and Scheduling Using Primavera® Version 4.1 or E&C
Planning and Scheduling Using Primavera® Version 5.0 – For IT Project Office
Planning and Scheduling Using Primavera® Version 5.0 – For Engineering & Construction
Project Planning & Control Using Primavera® P6 – Updated for Version 6.2
Planning Using Primavera Project Planner P3® Version 2.0
Planning Using Primavera Project Planner P3® Version 3.0
Planning Using Primavera Project Planner P3® Version 3.1
Project Planning Using SureTrak® for Windows Version 2.0
Planning Using Primavera SureTrak® Project Manager Version 3.0
Planning and Control Using Oracle Primavera P6 - Version 8.1 Professional Client & Optional Client
Planning & Control Using Primavera® P6™ For all industries including Versions 4 to 7
Planning and Control Using Oracle Primavera P6 - Versions 8.2 EPPM Web
Planning and Control Using Oracle Primavera P - Versions 8.2 & 8.3 EPPM Web
Planning and Control Using Oracle Primavera P6 - Version 8.1 & 8.2 Professional Client & Optional Client
Planning and Control Using Oracle Primavera P6 - Version 8.1, 8.2 & 8.3 Professional Client & Optional Client
Project Planning and Control Using Oracle Primavera P6 - Versions 8.1, 8.2 & 8.3 Professional Client & Optional Client
Planificación y Control de Proyectos Usando Oracle Primavera P6 - Versiones 8.1, 8.2 y 8.3 Cliente Profesional y Opcional
Planning and Control Using Microsoft Project 2010 & *PMBOK® Guide* Fifth Edition
项目规划和控制 ORACLE PRIMAVERA P6 应用 - 版本 8.1, 8.2 & 8.3 专业&可选客户端
Project Planning and Control Using Oracle Primavera P6 - Versions 8.1 to 8.4 Professional Client & Optional Client
Planning and Control Using Oracle Primavera P6 - Versions 8.2 to 8.4 EPPM Web
Planificación y Control de Proyectos Usando Oracle Primavera P6 Versiones 8.1 a 8.4 Cliente Profesional & Cliente Opcional
规划和控制 ORACLE® PRIMAVERA® P6 应用 版本 8.1-8.4 专业&可选客户端

© *Eastwood Harris Pty Ltd*

1 INTRODUCTION

1.1 Purpose

The purpose of this book is to provide you with a method for planning and controlling projects using Microsoft Office Project Professional 2013 and 2016, or Microsoft Office Project Standard 2013 and 2016, in a single project environment up to an intermediate level.

Microsoft Project 2016 is a minor upgrade with very few functional differences. One of the more noticeable differences is the issue that Microsoft Project 2016 has put large spaces between the buttons in Quick Access making the Quick Access Toolbar less useful than in earlier versions This book highlights the important differences between Microsoft Office Project 2013 Microsoft Project 2016 and earlier versions, but it is not intended to be used for learning these earlier versions.

The screen shots in this book were captured using Microsoft Office Project Standard and Professional 2013 & Windows 8, and the revised workshops and new functions in Microsoft Project 2016 were captured with Microsoft Project Standard 2016 and Windows 10.

Readers using Microsoft Project Professional will have more menu options than those using Microsoft Project Standard. These additional menu options will operate when a project is saved to a Microsoft Project Server. Microsoft Office Project Standard will not operate with Microsoft Project Server and has fewer functions than Microsoft Project Office Professional operating with Microsoft Project Server.

At the end of this book, you should be able to:

- Understand the steps required to create a project plan
- Set up the software
- Define calendars
- Add tasks and organize tasks
- Format the display
- Add logic and constraints
- Use Tables, Views and Filters
- Print reports
- Record and track progress
- Create and assign resources and understand the impact of task types and effort-driven tasks
- Analyze resource requirements and level a schedule
- Update resourced projects
- Customize the project options
- Understand the different techniques for scheduling.

The book does not cover every aspect of Microsoft Project, but it does cover the main features required to create and update un-resourced and resourced project schedules. There are some chapters at the end of the book covering some more advanced features of the software. It should provide you with a solid grounding, which will enable you to learn the other features of the software by experimenting with the software, using the Help files and reviewing other literature and forums.

This book has been written to minimize superfluous text, allowing the user to locate and understand the information contained in the book as quickly as possible. It does NOT cover functions of little value to common project scheduling requirements. If at any time you are unable to understand a topic in this book, it is suggested that you use the Microsoft Project Help menu to gain a further understanding of the subject.

This publication is only sold as a bound book, no parts may be reproduced by any means, e.g. electronic, video or print.

© **Eastwood Harris Pty Ltd**   1

The term "button" is used for an object on the screen that may be clicked with the mouse to access a software function and the term "icon" for something that conveys information.

1.2   Required Background Knowledge

The book is intended to teach you how to use Microsoft Project in a project environment. Therefore, to be able to follow this book you should have the following background knowledge:

- The ability to use a personal computer and understand the fundamentals of the operating system,

- Experience using application software such as Microsoft Office which would have given you exposure to the Windows menu systems and typical Windows functions such as copy and paste, and

- An understanding of how projects are managed, such as the processes that take place over the lifetime of a project.

1.3   Purpose of Planning

The ultimate purpose of planning is to build a model that allows you to predict which tasks and resources are critical to the timely completion of the project. Strategies may then be implemented to ensure that these tasks and resources are managed properly, thus ensuring that the project will be delivered both **On Time** and **Within Budget**.

Planning aims to:

- Identify the total scope of the project and plan to deliver it

- Evaluate different project delivery methods

- Identify products/deliverables required to deliver a project under a logical breakdown of the project

- Identify and optimize the use of resources and evaluate if target dates may be met

- Identify risks, plan to minimize them and set priorities

- Provide a baseline plan against which progress is measured

- Assist in stakeholders' communication, identifying what is to be done, when and by whom

- Assist management to think ahead and make informed decisions.

Planning helps to avoid or assists in evaluating:

- Increased project costs or reduction in scope and/or quality

- Additional project testing and handover costs

- Extensions of time claims

- Loss of your client's revenue

- Resolving contractual disputes with subcontractors and suppliers

- The loss of reputation of those involved in a project

- Loss of a facility or asset in the event of a total project failure.

1.4 Project Planning Metrics

Components normally measured and controlled using planning and scheduling software are:

- Scope
- Time
- Effort (resources)
- Cost.

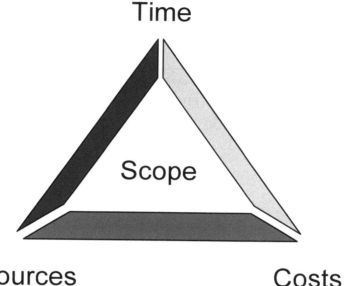

A change in any one of these components normally results in a change in one or more of the others.

Other project management functions that are not traditionally managed with planning and scheduling software, but may have components reflected in the schedule includes:

- Document Management and Control
- Quality Management
- Contract Management
- Issue Management
- Risk Management
- Industrial Relations
- Accounting.

The development of Enterprise Project Management systems has resulted in the inclusion of more of these functions in project planning and scheduling software.

1.5   Planning Cycle

The planning cycle is an integral part of managing a project. A software package such as Microsoft Project makes this task much easier.

When the original plan is agreed to, the **Baseline** or **Target** is set. The **Baseline** is a record of the original plan. The **Baseline** dates may be recorded in Microsoft Project in data fields titled **Baseline Start** and **Baseline Finish**.

After project planning has ended and project execution has begun, the actual progress is monitored, recorded in the software and compared to the **Baseline**.

The plan may be changed by adding or deleting tasks and adjusting Remaining Durations or Resources.

A revised plan is then published as progress continues.

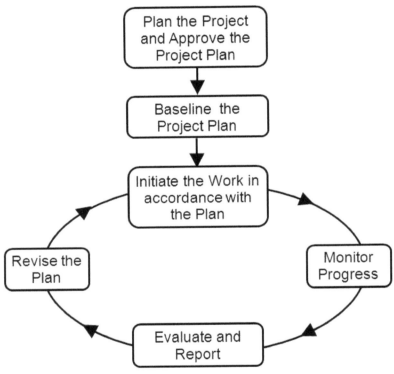

Updating a schedule assists in the management of a project by recording and displaying:

- Progress and the impact of project scope changes and delays as the project progresses,

- The revised completion date and final forecast of costs for the project,

- Historical data that may be used to support extension of time claims and dispute resolution, and

- Historical data that may be used in future projects of a similar nature.

1.6   Levels of Planning

Projects are often planned at a summary level and then detailed out in the schedule before the work is commenced. Smaller projects may be detailed out during project planning but other large or complex projects may require several levels before the project plan is fully detailed out.

The main reason for not detailing out a project early is that there may not be enough information at that stage; and time would be wasted in preparing detailed schedules that will be made redundant by unforeseen changes. The following planning techniques may be considered:

The PMBOK® Guide discusses the following techniques:

- The **Rolling Wave**. This technique involves adding more detail to the schedule as the work approaches. This is often possible, as more information is known about the scope of the project as work is executed. The initial planning could be completed at a high level in the **Work Breakdown Structure** (**WBS**). As the work approaches the planning may be completed in more detail at a lower level such as **WBS Component** and then to a **Work Package** level for detailed planning.

- The use of **Sub-projects**. These are useful in larger projects where more than one person is working on the project schedule. This situation may exist when portions of projects are contracted out. A sub-project may be detailed out when the work is awarded to a contractor.

- The use of **Phases**. A Phase is different from a PRINCE2 Stage as Phases may overlap in time and Stages do not. Phases may be defined, for example, as Design, Procure and Install. These Phases may overlap, as Procurement may commence before Design is complete. The Phase development of a schedule involves the detailing out of all the associated WBS elements prior to the commencement of that Phase.

- The PMBOK® Guide does not have strict definitions for levels of plans but assumes that this process is planned when decomposing the **Work Breakdown Structure** (**WBS**). There are some other models available that may be used as guidelines, such as the PMI "Practice Standard for Work Breakdown Structures."

- Projects subject to many changes may require the work to be granulated out into more tasks immediately prior to the work starting. For example, at the end of every week the next two weeks' work (that will have had been planned at a summary level) would be planned out in detail. This is a type of **Rolling Wave** development of the project. This is often found in a software-testing phase, where the next portion of work is dependent on what has passed and what has failed testing in the previous portion of work.

This publication is only sold as a bound book, no parts may be reproduced by any means, e.g. electronic, video or print.

© **Eastwood Harris Pty Ltd**   5

PRINCE2 Level of Plans

PRINCE2 is a project management methodology that was developed in the UK. This methodology defines the type of plans that a project team should consider.

Stages in PRINCE2 are defined as time-bound periods of a project, which do not overlap in time and are referred to as Management Stages. Under PRINCE2, a Project Plan is divided into Stages and a Stage plan is detailed out prior to the commencement of a Stage. PRINCE2 defines the following levels of plans:

- **Programme Plan** – which may include Project Plans or one or more portfolios of multiple projects,

- **Project Plan** – this is mandatory and is updated through the duration of a project,

- **Stage Plan** – there is a minimum of two Stage Plans, an **Initiation Stage Plan** and **First Stage Plan**. There would usually be one Stage Plan for each Stage.

- **Exception Plan** – which is at the same level and detail as a Stage Plan and replaces a Stage Plan at the request of a Project Board when a Stage is forecast to exceed tolerances (contingent time), and

- **Team Plan** – which is optional and would be used on larger projects where Teams are used for delivering Products that require detailed planning. A typical example is a contractor's plan, which would be submitted during the bid process.

Jelen's Cost and Optimization Engineering

The book outlines several levels of planning, each of which goes deeper in the planning process, depending on the level of granularity the planner wants to achieve:

- Level 0: Corresponds to the project as a whole, therefore it will show as an individual bar in the Gantt Chart which extent will be from the start to the finish date of the project.

- Level 1: This level programmes the project according to its most significant parts; hence, it will show like a Bar Chart on the Gantt Chart. For example, visualize a schedule for a food factory, which could be broken down into quality control, processing, packaging and storage.

- Level 2: This level takes the previous one as a foundation to keep splitting down the parts of the project. Generally the schedule will be shown in a bar chart format and it could have some constraints against it. Following the previous example, the storage area in the food factory could be further divided into dried foods, fresh produce, canned products, etc.

- Level 3: The breakdown goes even further and at this point it becomes very clear where the critical path is. Likewise, since this division is not too detailed, the level of control over the project can be performed very well.

- Levels 4 – n: From this level, is up to the planner's decision to keep breaking down the project into more detailed parts. This type of planning is usually performed to build a look-ahead of the schedule between 1 and 6 months. The most common practice is to show these schedules in a Bar Chart format or as CPMs.

1.7   Monitoring and Controlling a Project

After a plan has been produced, it should be followed and the work authorized in accordance with the plan. If there is to be a change in the plan, then the plan should be formally changed. If necessary, the client should be informed and when required by the contract, approval for the change should be sought. If the plan is not followed, then it may be difficult to obtain approval for extension of time claims and will make dispute resolution more difficult.

Monitoring a project ensures the project provides management with historical data, trends and deviations from plan allowing them to make decisions and control the project. It should record the:

- Dates that work was started and completed

- Hours and cost required to complete the work

- Deliverables/products produced

Monitoring would also:

- Confirm the required quality is being met

- Provide trends that may be used for forecasting the remainder of the project

- Record historical data for use in assisting in planning future projects

- Obtain data required for preparing extension of time claims and for dispute resolution.

Controlling is the next level of management, analyzing the progress to date and trends to:

- Ensure that the project is being executed according to the plan

- Compare the project's progress with the original plan

- Review options in the case of deviations

- Forecast problems as early as possible, which enables corrective action to be taken as early as possible.

2   CREATING A PROJECT SCHEDULE

2.1   Understanding Planning and Scheduling Software

The aim of this chapter is to give you an understanding of what a plan is and some practical guidance on how your schedule may be created and updated as part of a project.

The project is essentially a set of operations or tasks required to be completed in a logical order to provide one or more deliverables or products. A schedule is an attempt to model these operations and their relationships. These operations take time to accomplish and may employ resources that have limited availability.

Planning and scheduling software allows the user to:

- Enter the hierarchical structure of the project deliverables or products into the software. This is often called a Work Breakdown Structure (WBS) or Product Breakdown Structure (PBS),

- Break a project down into tasks required to create the deliverables. These are entered into the software as **Tasks** under the appropriate WBS,

- Assign durations, constraints, predecessors and successors of the tasks and then calculate the start and finish date of all the tasks,

- Assign resources and/or costs, which represent people, equipment or materials, to the tasks and calculate the project resource requirements and/or cash flow,

- Optimize the project plan,

- Set Baseline Dates and Budgets to compare progress against,

- The plan should be used to approve the commencement of work,

- Record the actual progress of tasks against the original plan and amend the plan when required allowing for scope changes etc.,

- Record the consumption of resources and/or costs and re-estimate the resources and/or costs required to finish the project, and

- Produce management reports.

There are four modes or levels in which planning and scheduling software may be used.

	Planning	Tracking
Without Resources	LEVEL 1 Planning without Resources	LEVEL 2 Tracking progress without Resources
With Resources and/or Costs	LEVEL 3 Planning with Resources	LEVEL 4 Tracking progress with Resources

As the level increases, the amount of information required to maintain the schedule will increase and, more importantly, your skill and knowledge in using the software will also have to increase. This book is designed to take you from Level 1 to Level 4.

2.2 Understanding Your Project

Before you start the process of creating a project plan, it is important to have an understanding of the project and how it is to be executed. On large complex projects, this information is usually available from the following types of documents:

- Project charter or business case

- Project scope

- Functional specification

- Requirements baseline

- Contract documentation

- Plans and drawings

- Project execution plan

- Contracting and purchasing plan

- Equipment lists

- Installation and testing plan

- Historical data.

Many project managers conduct a **Stakeholder Analysis** at the start of a project. This process lists all the people and organizations with an interest in the project and their interests.

- Key project success factors may be identified from the interests of the most influential stakeholders.

- It is important to use the stakeholder analysis to identify all the stakeholders' tasks. These must be included in the schedule.

It is important to gain a good understanding of how the project is to be executed before entering any data into the software. It is considered good practice to plan a project before creating a schedule in any planning and scheduling software. These documents are referred to by many terms such as Project Execution Plan and Project Methodology Statement. You should also understand what level of reporting is required by the project team, as providing either too little or too much detail will often lead to the schedule being discarded.

The following processes are required to create or maintain a plan:

- Collecting the relevant project data,

- Entering and manipulating the data in software,

- Reviewing, revising and distributing the plan.

The ability to collect the data is as important as the ability to enter and manipulate the information into the software. On larger projects it may be necessary to write policies and procedures to ensure accurate collection of data from the various departments and sites.

The structure, level of detail in a schedule and how resources are created and assigned depends on the schedule purpose. Types of schedule include tender, forecasting cash flows, Critical Path contract master schedule & two week look-ahead for tasking people. They will all have a different structure, level of detail and resourcing philosophy.

2.3   Level 1 – Planning without Resources

This is the simplest mode of planning, but this has limitations and may create an unachievable schedule as resources will not be considered.

2.3.1   Creating Projects

To create the project you will require the following information:

- Project Name,

- Client Name,

- Other information such as Location

- The Start Date (or perhaps the Finish Date).

2.3.2   Defining Calendars

Before you start entering tasks into your schedule it is advisable to set up the calendars. They are used to model the working time for each task in the project. For example, a 6-day per week calendar is created for those tasks that will be worked for 6 days a week. The calendars should include any public holidays and any other exceptions to available working days such as Rostered Days Off (RDO).

The finish date and time of a task (when there are no resources using a resource calendar assigned to a task) is calculated from the start date and time plus the task duration over the calendar assigned to the task.

The pictures below show the effect of nonworking days on the finish date of a 13-day duration task assigned a 5-day working week. The elapsed duration is 19 days due to the three weekends when work does not take place.

February 2019						
M	**T**	**W**	**Th**	**F**	**S**	**S**
			1	2	3
4	5	6	7	8	9	10
11	12	13	14	15	16	17
18	19	20	21	22	23	24
25	26	27	28			

Duration	T	F	S	S	M	T	W	T	F	S	S	M	T	W	T	F	S	S	M	T	W	T
					11 February							18 February							25 February			
13 days																						

2.3.3 Defining the Project WBS Using Outlining

A WBS may be defined as a hierarchical breakdown of all the project deliverables or products. The sum of all the WBS Components (Nodes) should encompass the total project scope.

The principal method of assigning a WBS to a project with Microsoft Project is using a function entitled **Outlining**, which creates a hierarchy of summary tasks. The picture below shows a simple project where the total project duration and three WBS Nodes are represented by summary tasks and bars:

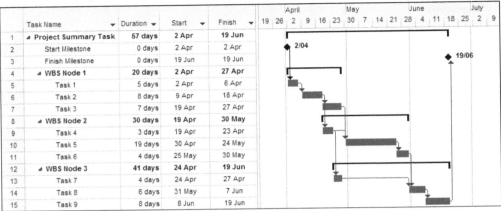

The project schedule may be rolled up using these summary tasks:

Microsoft Project also has customizable fields that may be used to assign codes and descriptions to tasks. The customizable fields may be used to sort, select, summarize and group tasks under codes and identify the project breakdown structure, thus providing a lot more flexibility than the Outlining function for developing and reporting a project schedule.

Before creating a WBS the following questions should be considered:

- What are the deliverables of the project, what is to be handed to the client/customer?

- What needs to be created to enable the final deliverable to be handed over to the client/customer?

- How many phases are there? (e.g., Design, Procure, Install and Test)

- How many disciplines are there? (e.g., Civil, Mechanical and Electrical)

- Which departments are involved in the project? (e.g., Sales, Procurement and Installation)

- How many locations, buildings or floors are involved in the project?

- What work is expected to be contracted?

- How many sites or areas are there in the project?

Use the responses to these questions to create the WBS data dictionaries.

As the project progresses it may be necessary to add more detail to the project and additional WBS Nodes may be added as the scope develops or more detail is added to the schedule.

2.3.4 Defining, Adding and Organizing Tasks

Activities must be defined before they are entered into the schedule. It is important that you carefully consider the following factors:

- What is the scope of the activity? (What is included and excluded?)

- How long is the activity going to take?

- Who is going to perform it?

- What are the deliverables or output for each activity?

The project estimate is usually a good place to start looking for a breakdown of the project into activities, resources, and costs. It may even provide an indication of how long the work will take.

Activities may have variable durations depending on the number of resources assigned. You may find that one activity that takes 4 days using 4 workers may take 2 days using 8 workers or 8 days using 2 workers.

Usually project reports are issued on a regular basis such as every week or every month. It is recommended that, if possible, an activity should not span more than two reporting periods. That way the activities should only be **In-progress** for one report. Of course, it is not practical to do this on long duration activities, such as procurement and delivery activities, that may span many reporting periods.

Good practice recommends that you have a measurable finish point for each group of activities. These may be identified in the schedule by **Milestones** and are designated with zero duration. You may issue documentation to officially highlight the end of one activity and the start of another, thereby adding clarity to the schedule. Examples of typical documents that may be issued for clarity are:

- Issue of a drawing package

- Completion of a specification

- Placing of an order

- Receipt of materials (delivery logs or tickets or dockets)

- Completed testing certificates for equipment or systems

2.3.5 Adding the Logic Links

The logic is added to the schedule to provide the order in which the tasks must be undertaken. The logic is designated by indicating the predecessors to, and the successors from, each task. There are two methods that software uses to sequence tasks:

- Precedence Diagram Method (PDM), where the node is the task and the arrow is the relationship, and

- Arrow Diagram (ADM), where the node is the relationship and the arrow is the task.

Most current project planning and scheduling software uses PDM, including Microsoft Project. A PDM diagram may be created using the Network Diagram view.

There are several types of dependencies that may be used:

- **Mandatory dependencies**, also known as **Hard Logic**, are relationships between tasks that may not be broken. For example, a hole has to be dug before it is filled with concrete or a computer delivered before software is installed onto it.

- **Discretionary dependencies**, also known as **Sequencing Logic** or **Soft Logic**, are relationships between tasks that may be changed when the plan is changed. For example, if there are five holes to be excavated and only one machine available, or five computers to be assembled and one person available to work on them, then the order of these tasks could be set with sequencing logic but changed at a later date.

Both **Mandatory dependencies** and **Discretionary dependencies** are entered into Microsoft Project as task relationships or logic links. The software does not provide a method of identifying the type of relationship because notes or codes may not be attached to relationships. A **Note** may be added to either the predecessor or successor task to explain the relationship.

External dependencies are usually events outside the control of the project team that impact on the schedule. An example would be the availability of a site to start work. This is usually represented in Microsoft Project by a Milestone which has a constraint applied to it. This topic is discussed in more detail in the next section.

The software will calculate the start and finish dates for each task. The end date of the project is calculated from the start date of the project, the logic among the tasks, any **Leads** (often referred to as **Negative Lag**) or **Lags** applied to the logic and durations of the tasks. The pictures below show the effect of a lag and a lead on the start of a successor task:

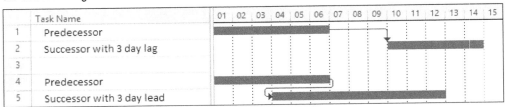

It is good practice to create a **Closed Network** with the logic. A **Closed Network** is created when all tasks have one or more start predecessors and one or more finish successors, except:

- One or more project start milestones or incoming External Dependencies which have no predecessors, and

- One or more finish milestones or outgoing External Dependencies which have no successors.

The project's logic must not loop back on itself. Looping would occur if the logic stated that A preceded B, B preceded C, and C preceded A. That's not a logical project situation and will cause an error comment to be generated by the software during network calculations.

Thus, when the logic is correctly applied, a delay to a task will delay all its successor tasks and delay the project end date when there is insufficient slippage time to accommodate the delay. This spare time is usually called **Float** (but Microsoft Project uses the term **Slack**).

2.3.6 Developing a Closed Network

It is good practice to create a **Closed Network** with the logic. In a **Closed Network**, all activities have one or more predecessors and one or more successors except:

- The project start milestone or first activity, which has no predecessors, and

- The finish milestone or finish activity, which has no successors.

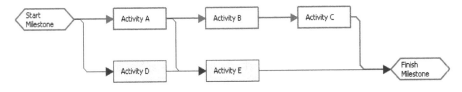

 When a closed network is not established then the Critical Path, Total Float and Free Float will not calculate correctly

The project's logic must not loop back on itself. Looping occurs if the logic states that A preceded B, B preceded C, and C preceded A. That is not a logical project situation and will cause the software to generate an error comment during network calculations.

Thus, when the logic is correctly applied, a delay to an activity will delay all its successor activities and delay the project end date when there is insufficient spare slippage time to accommodate the delay. This spare time is normally called **Float** but note that Microsoft Project uses the term **Slack** for **Float**.

2.3.7 Constraints

External dependencies are applied to a schedule using **Constraints** and these may model the impact of events outside the logical sequence of tasks. A constraint would be imposed to specific dates such as the availability of a facility to start work or the required completion date of a project. Constraints should be cross-referenced to the supporting documentation such as Milestone Dates from contract documentation using the **Notes** function. Typical examples of constraints would be:

- **Start No Earlier Than** for the availability of a site to commence work, and

- **Finish No Later Than** for the date that a total project must be delivered by or handed over to a client.

2.3.8 Risk Analysis

The process of planning a project may identify risks and a formal risk analysis should be considered. A risk analysis may identify risk mitigation tasks that should be added to the schedule before it is submitted for approval.

2.3.9 Contingent Time

The addition of contingent time should be considered when submitting a schedule for approval. Estimates usually have contingency and if this money is to be expended then an allowance for time to spend the contingent funds needs to be made. Contingent time may be assigned to a schedule in a number of ways including:

- One or more contingent time tasks could be inserted in the project. These would be adjusted in length as the project progresses to maintain the planned end date.

- Work days in the calendar could be assigned as nonworking days. For example a building project could be scheduled on a 5-day per week basis, knowing that work will be undertaken on the Saturday.

- Increasing the task durations by a factor.

- Positive lags could be assigned between tasks, but this is not recommended by the author.

A lack of contingency should be identified as a Risk.

2.3.10 Printing and Reports

There are software features that enable you to present the information in a clear and concise manner to communicate the requirements to all project members. These functions are covered in the **Printing and Reports** chapter.

© *Eastwood Harris Pty Ltd*   15

2.3.11 Issuing the Plan

All members of the project team should review the project plan in an attempt to:

- Optimize the process and methods employed, and

- Gain consensus among team members as to the project's logic, durations, and Project Breakdown Structures.

Team members should communicate frequently with each other about their expectations of the project while providing each with the opportunity to contribute to the schedule and further improve the outcome.

2.3.12   Scheduling the Project and Understanding Float (Slack)

The software will calculate the shortest time in which the project may be completed.

It will also identify the **Critical Path(s)**. The Critical Path is the chain(s) of tasks that takes the longest time to accomplish. This will define the Earliest Finish date of the project. The calculated completion date depends on the critical tasks starting and finishing on time – if they are delayed, the whole project will be delayed.

Thus, when the logic is correctly applied to both relationships and constraints, a delay to a task will delay all successor tasks, and, in turn, postpone the project end date. When there is insufficient spare time or **Float** to accommodate the delay. Tasks that may be delayed without affecting the project end date have **Float**.

 Float is the term that is in common use and one that most textbooks refer to. Microsoft Project uses the term **Slack** and not **Float**. Either term may be used.

Total Float is the amount of time a task may be delayed without delaying the project end date. The project in the picture below is on a 5-day per week calendar, thus there is no work over the weekends. Task 4 and Task 5 both have Total Float. The delay of a task with Total Float may delay other tasks with Total Float. Task 4 would have to delay Task 5 before a delay to Task 4 would delay the project Finish Milestone.

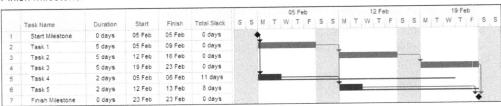

Free Float is the amount of time a task may be delayed without delaying the start date of another task. Task 4 has only 3 days' Free Float because once it has been delayed 3 days it will delay Task 5.

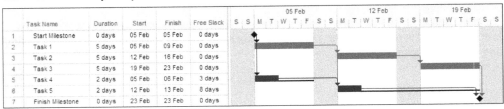

2.3.13 Formatting the Display – Filters and Views

Now it is time to present the schedule for approval. There are tools to manipulate and display the tasks to suit the project reporting requirements:

- **Filters** reduce the number of tasks that are displayed
- **Tables** display the data in columns, and
- **Views** allow the presentation of the columns of data and bars to be tailored.

2.3.14 Printing and Reports and Issuing the Plan

There are features that allow you to present the information in a clear and concise manner to communicate the requirements to all project members.

All members of the project team should review the project plan in an attempt to optimize the process and methods employed.

Printouts and reports should be used to communicate what is expected of team members while providing each with the opportunity to further improve the outcome.

2.4 Level 2 – Tracking Progress without Resources

2.4.1 Setting the Baseline Schedule

The optimized and agreed-upon plan is used as a baseline for future comparisons. The software can record the original planned dates, which are often called Baseline dates, for comparing the actual progress to the plan over the life of the project. These planned dates are stored in the **Baseline Date** fields and displayed as the upper bars in the picture below, which demonstrates that there has been a delay in the start of the project:

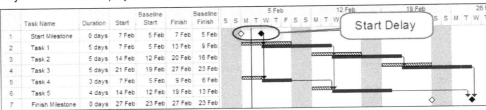

2.4.2 Tracking Progress

The schedule should be **Updated** (or progressed) on a regular basis and progress recorded at that point. The date on which progress is reported is known by a number of different terms such as **Data Date**, **Update Date**, **Time Now**, **Report Date** and **Status Date**. The Status Date is the field used in Microsoft Project to record this date. Whatever the frequency chosen for updating, you will have to collect the following task information in order to update a schedule:

- Completed tasks
  - ➢ Actual Start date and
  - ➢ Actual Finish dates.
- In-progress tasks
  - ➢ The Actual Start Date
  - ➢ Percentage Completed
  - ➢ The Remaining Duration from the Status Date or Expected Finish Date of the Task
  - ➢ Changes to logic.
- Un-started work
  - ➢ Any revisions to tasks that have not started
  - ➢ New tasks representing scope changes
  - ➢ Forecast start date for tasks due to start in the near term
  - ➢ Revisions to logic that represent changes to the plan.

The schedule may be updated after this information has been collected. The recorded progress is compared to the **Target** dates, either graphically or by using columns of data such as the **Finish Variance** column:

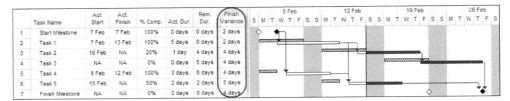

2.4.3 Corrective Action

At this point it may be necessary to further optimize the schedule to bring the project back on track. Possible options include:

- Reduce the contingent time allowance.

- Assign a negative lag on a Finish-to-Start relationship allowing a successor to commence before a predecessor is completed.

- Change relationships to Start-to-Start allowing tasks to be executed in parallel.

- Reduce the durations of tasks. In a resourced schedule this could be achieved by increasing the number of resources assigned to a task.

- Work longer hours per day or days per week by editing the calendars, or

- Reduce the scope and delete tasks.

2.5   Level 3 – Planning with Resources and /or Costs

2.5.1   Estimating or Planning for Control

There are two modes that the software may be used at Level 3.

- **Estimating**. In this mode the objective is to create a schedule with costs that are being used as an estimate so the schedule will never be updated. Tasks may have many resources assigned to them to develop an accurate cost estimate and include many items that would never be updated in the process of updating a schedule.

- **Planning for Control**. In this mode the intention is to assign resource actual units (hours) and costs, then calculate units and costs to completion. It may also be intended to conduct an Earned Value analysis. In this situation it is important to ensure the minimum number of resources are assigned to tasks, and preferably only one resource assigned to each task. The process of updating a schedule becomes extremely difficult and time consuming when a resourced schedule has many resources per task. The scheduler is then in danger of becoming a timekeeper and may lose sight of other important functions, such as calculating the forecast to complete and the project finish date.

2.5.2   The Balance Between the Number of Tasks and Resources

When planning to control a large or complex project, it is important to maintain a balance between the number of tasks and the number of resources that are planned and tracked. As a general rule, the more tasks a schedule has, the fewer resources should be created and assigned to tasks.

When there is a schedule with a large number of tasks and a large number of resources assigned to each task, the schedule may end up in a situation where no members of the project team are able to understand the schedule and the scheduler is unable to maintain it.

Instead of assigning individual resources such as people by name, consider using generic "Skills" or "Trades," and on very large projects use "Crews" or "Teams."

Therefore it is more important to minimize the number of resources in large schedules that will be updated regularly, since updating every resource assigned to each task at each schedule update is very time consuming.

2.5.3   Creating and Using Resources

A resource pool is established by entering the project resources and their unit rates into the software. Entering a cost rate for each resource enables cost analysis such as comparing the cost of supplementing overloaded resources against the cost of extending the project deadline.

Resources are then assigned to tasks and the time-phased resource requirements, cash flows and budgets may be automatically produced from this resource/cost data.

2.5.4   Resource Calendars, Task Types and Driving Resources

These are additional features that enable the user to more accurately model real-life situations. These features add a level of complexity that should be used only when the environment demands their use and should be avoided by inexperienced schedulers.

2.5.5   Resource Graphs and Usage Tables

These features allow the display and analysis of project resource requirements in both tables and graphs.

The data may be exported to Excel for further analysis and presentation.

2.5.6   Resource Optimization

The schedule may now have to be resource optimized to:

- Reduce peaks and smooth the resource requirements, or

- Reduce resource demand to the available number of resources, or

- Reduce demand to an available cash flow when a project is being financed on a customer's income, or

- Efficiently utilize resources.

The process of leveling is defined as delaying tasks until resources become available. There are several methods of delaying tasks and thus leveling a schedule, which are outlined in the **RESOURCE OPTIMIZATION** chapter.

2.6   Level 4 – Tracking Progress of a Resourced Schedule

2.6.1   Updating Projects with Resources

When you update a project with resources you will need to collect some additional information that may include:

- The quantities or hours and/or costs spent to date per task for each resource, and

- The quantities or hours and/or costs required per resource to complete each task.

You may then update a resourced schedule with this data.

Once the schedule has been updated then a review of the future resource requirements, project end date and cash flows may be made.

Updating a resourced project is time consuming and requires experience and a good understanding of how the software calculates. It should ideally be attempted by experienced schedulers or under the guidance of an experienced scheduler.

3   NAVIGATION AND SETTING THE OPTIONS MICROSOFT PROJECT

3.1   *Starting Microsoft Project Professional and Standard*

When you start Microsoft Project for the first time and/or after going through the licensing process you may be shown a screen like the one below;

The start-up screens are slightly different for the Standard and Professional versions because the Professional has more functions and is able to connect to a Microsoft Project Server and open files from a server.

The buttons on the right allow you to create new projects from a variety of self-explanatory sources.

You may:

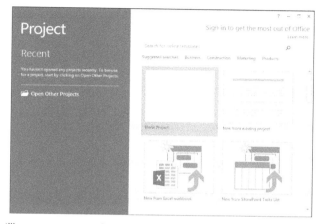

- Create a **Blank Project,** which utilizes your system **Global.mpt** template,

- Creating a project from a Personal Template, but you will need to set your template directory and save a template first, this topic is covered over the page and in detail in para 4.5

- Copy an existing project, by using the **New from existing project** option,

- Import data from an **Excel** spreadsheet, by using the **New from Excel workbook**,

- Import a task list from **SharePoint**, by using the **New from SharePoint Task list**,

- Create a project from one of the Microsoft templates, when connected to the internet, by using the **Search for online templates** option,

- The **Get Started** button is a tutorial to guide you through how to create a project.

The **Open Other Projects** option takes you to the **Open** screen:

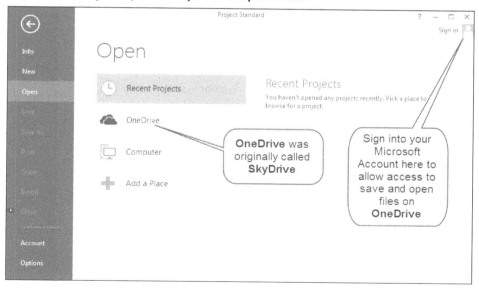

Unlike earlier versions of Microsoft Project, a blank project titled **Project 1** is not automatically created with a default load of Microsoft Project 2013 or 2016, so you will either need to:

- Create a new project or open an existing project before you may start working, or

- If you wish the software to start up in the Gantt Chart View with a blank project titled **Project 1** created (as in earlier version of Microsoft Project) then you will need to select **File**, **Options**, **General**, **Start up options** and uncheck the **Show the Start screen when this application starts**:

The **Open** screen has options to:

- **Recent Projects** allows you to open a recently opened project, they will be listed here when you have opened projects before,

- Open a project from **OneDrive** which a Microsoft facility like **DropBox** that allows you to access files from anywhere and share them with other people. This was formally called **SkyDrive**, as shown in the picture below.

- **Computer** allows you access to drives on your computer to open a project from a drive on your computer.

- **Add a Place** allows you to map other locations for quicker access to where your files are stored.

- **Sign in** at the top right hand side of the scree allows you to sign in and access your Microsoft account so you may open and save files to **OneDrive** accounts.

The other menu options on left hand side are:

- **New** takes you to a screen to create new projects.

- **Account** takes you to a screen to log into your Microsoft Account and once you are signed in you may open and save files to the Microsoft **OneDrive** (**SkyDrive**).

- **Options** allows you to change the options which we discuss later.

Once a new project has been created there will be more options available from the **File** menu:

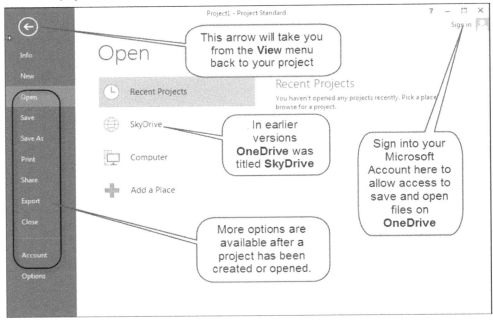

3.2 Identify the Parts of the Project Screen

After starting Microsoft Project and creating a new blank project, the name for the file will be titled **Project1** and the default Microsoft Project screen will look like this:

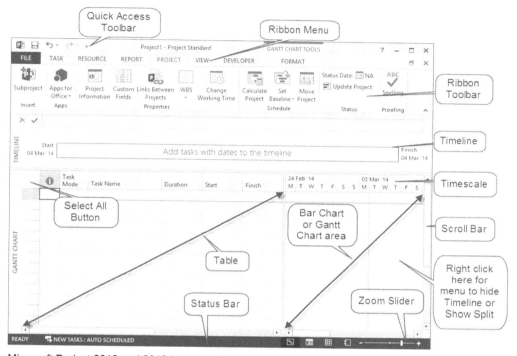

Microsoft Project 2013 and 2016 have the **Ribbon** style menu system with a typical Windows look-and-feel.

- The **Ribbon** works in a similar way to the Microsoft Office products with a Ribbon.
- The project name is displayed before **Microsoft Project** at the top left-hand side of the screen.
- The **Ribbon** by default is displayed below the **Quick Access Toolbar**. This is a combination of both the traditional menu and toolbar buttons.
- The main display has the **Bar Chart** or **Gantt Chart** on the right-hand side, the **Timescale** above, and the **Table** of Data Columns on the left-hand side with their titles above them. The divider between the two areas may be dragged from side to side by holding down the left mouse button.
- The horizontal **Scroll Bars** are at the bottom of the screen and the **Status Bar** is below the **Scroll Bars**. The vertical **Scroll Bar** is at the right-hand side of the screen.
- The new **Zoom Slider** which is used for scaling the timescale is displayed in the bottom right-hand side of the screen in the **Status Bar**.

 The grouping of the Ribbon commands is by software functions such as Project, Task etc. and not by scheduling functions such as creating or updating. Users therefore continually swap from one **Ribbon** menu to another while operating the software. You should therefore consider hiding the **Ribbon** and building your own **Quick Access Toolbar** below the **Ribbon** with the commands in logical groupings or downloading the Eastwood Harris Quick Access toolbar from www.eh.com.au Software Downloads page.

3.3 Customizing the Screen

The screen may be customized in a number of ways to suit your preferences. With the implementation of the Ribbon Toolbar introduced in Microsoft Project 2010 users migrating from earlier versions of Microsoft Project will have to learn the new location of all the commands and new pictures for many command buttons.

3.3.1 Ribbon Toolbar

The toolbars will not be covered in detail in this book as they operate the same way as all other Microsoft products but the key issues will be outlined.

The Ribbon Toolbar has **Tabs** along the top and **Ribbon Groups**, which are groups of **Command Buttons**, below the Ribbon Tabs. These Ribbon Groups were either menu items or toolbars in earlier versions of Microsoft Project.

The Ribbon Format tab Groups of buttons will change as different Views are opened. Two examples of different Formats Groups of buttons are displayed below:

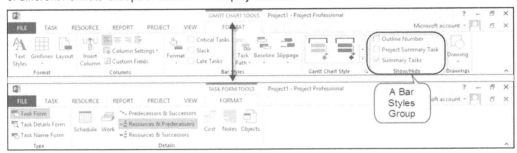

Significant productivity improvements may be made by ensuring that frequently used functions are made available on the **Quick Access Toolbar** and right-clicking in the **Quick Access Toolbar** will display a menu allowing customizing of the **Quick Access Toolbar**, see the picture below:

- The **Quick Access Toolbar** should be moved below the **Ribbon Toolbar** by clicking on the **Show Quick Access Toolbar Below the Ribbon**. It is recommended that you select this option and move the Quick Access Toolbar below the Ribbon Toolbar as more buttons may then be displayed and the bar is not truncated by the Project Name at the top of the screen.

- The **Add to Quick Access Toolbar** command will add a selected Ribbon Toolbar button to the Quick Access Toolbar. By default the Quick Access Toolbar is positioned at the top left-hand side of the screen and is always displayed.

- The **Customize Quick Access Toolbar…** opens the **Project Options** form **Quick Access Toolbar** tab where buttons may be added to or removed from the Quick Access Toolbar.

- **Customize the Ribbon…** opens the **Project Options** form, **Customize Ribbon** tab where buttons may be added to or removed from the Ribbon Toolbar.

- **Collapse the Ribbon** hides the Ribbon Toolbar and just leaves the menu displayed at the top. This was called Minimize the Ribbon in Microsoft Project 2010. Clicking on one of the menu items will display the Ribbon Toolbar. This is a recommended setting.

- The Ribbon Toolbar may also be minimized by clicking on the ⌃ buttons in the top right-hand corner of the screen.

Ribbon Tabs, **Groups** of command buttons and individual **Command Buttons** may be added to or removed from the **Ribbon** by selecting **File**, **Options** and selecting the **Customize Ribbon** tab.

Some Ribbon tabs will change their contents to suit the view that has been selected and is active; this is covered further in the next paragraphs.

3.3.2 Quick Access Toolbar

One of the more noticeable changes in the release of Microsoft Project 2016 is that it now has put large spaces between the buttons in Quick Access Toolbar, making the Quick Access Toolbar less useful than in earlier versions. In the two pictures below you may see that there are many more Quick Access Toolbar buttons in the Microsoft Project 2013 screenshot than the Microsoft Project 2016 screenshot:

Microsoft Project 2016

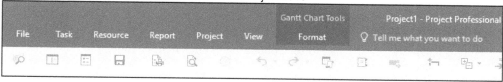

Microsoft Project 2013

In the picture below, which is from Microsoft Project 2013 the:

- Quick Access Toolbar may be moved below the Ribbon bar by clicking on the **Show Quick Access Toolbar Below the Ribbon**. **THIS IS STRONGLY RECOMMENDED SO THE MAXIMUM NUMBER OF COMMANDS ARE ALWAYS VISIBLE.**

- The Ribbon Toolbar has been hidden using **Collapse the Ribbon**. **THIS IS ALSO STRONGLY RECOMMENDED TO MAXIMIZE THE SIZE OF THE SCREEN.**

Click on the button on the **Quick Access Toolbar** to display the Quick Access Toolbar menu:

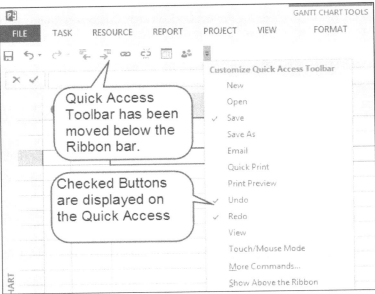

- The buttons that are checked are displayed in the Quick Access Toolbar.

- Buttons left unchecked will be displayed in the drop-down menu by clicking on the **Customize Quick Access Toolbar** button.

- **More Commands…** opens the **Project Options** form, **Quick Access Toolbar** tab where buttons may be added to or removed from the Ribbon Toolbar.

3.3.3   Exporting and Importing Toolbars

Both the Ribbon and Quick Access toolbars may be exported and imported from one computer to another and are compatible with both Microsoft Project 2010, 2013 and 2016, allowing users to transfer their customized tool bars to other computers they use, or to import toolbars created by other users. To import and export toolbars:

- Select **File**, **Options** and select **Customize Ribbon** or **Quick Access Toolbar**, and

- Use the **Import/Export** option at the bottom right hand corner of the form.

i You should consider downloading and importing the Eastwood Harris Quick Access toolbar from www.eh.com.au, **Software Downloads** page. This toolbar has many of the frequently used functions added and grouped into logical groups.

3.4   Microsoft Project Windows

3.4.1   Understanding Windows in Microsoft Project

In Microsoft Project multiple projects may be opened and multiple windows for each project may also be opened at the same time. Each window may be assigned a different View. This is similar to the way Asta Powerproject operates.

The picture below shows three views of the same project that have been opened at the same time, with the Gantt Chart view on the left being split:

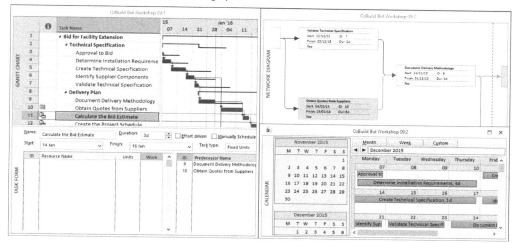

3.4.2   Creating a New Window

As each project is opened it is displayed with the Gantt Chart View. Additional windows may be created by:

- Selecting **View**, **Window** group, **New Window**,

- Select the project that requires the new window from the **New Window** form,

- The new window may be formatted using the **Ribbon**, **View** tab. This topic is covered in later chapters.

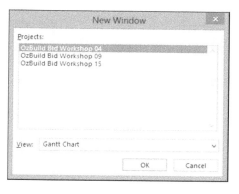

3.4.3   Managing Windows

The **Ribbon**, **View** tab, **Window** group has the following **Command Buttons**:

- **Switch Windows** 🔲 – allowing another window to be selected,

- **Arrange All** 🔲 – arranging the windows as per the picture above,

- **Hide Window** 🔲 – which will hide a window from view, the window will not be displayed using the **Arrange All** function and the **Hide /Unhide** command has to be used to display the window again.

3.4.4   Resizing Windows

A window may be:

- **Resized** by dragging with the mouse, or

- **Maximized** using the **Maximize** button at the top right-hand corner of each window.

3.4.5   Splitting Views

By default, the **Timeline** view is displayed above the Gantt Chart, this may be hidden by right clicking in the Bar Chart area. A window may be split horizontally into two panes. A different **View** may be displayed in each pane. This is termed **Dual-Pane view**. To open or close the dual-pane view:

- Select **View**, **Split View** group **Details** check box to split or uncheck to remove the split, or

- Grab the horizontal dividing bar at the bottom of the screen by holding down the left mouse button and dragging the line to resize the panes.

- Right-click in the right-hand side of the top pane and you will, in most views, be able to display a menu with a check box to **Show Split**.

- Double-click the dividing line or drag it to also remove or open the split window.

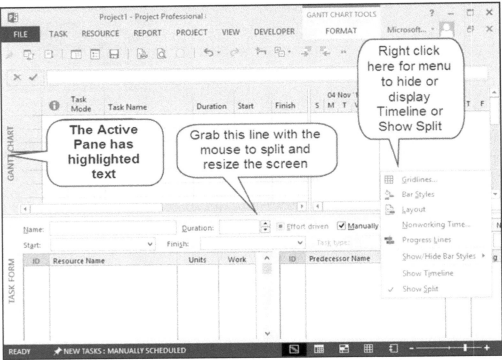

A pane needs to be **Active** before menu items pertaining to that pane become available. In earlier versions of Microsoft Project a dark band on the left-hand side displayed the **Active Pane**. In Microsoft Project 2013 and 2016 there is a change to the font highlighting to show the active pane which is not as obvious.

The **Ribbon**, **View** tab options will often change when different **Views** are selected in a pane.

A **Pane** is made **Active** just by:

- Clicking anywhere in the pane, or
- Pressing **F6** to swap active panes.

 Only the **Active Pane** may be printed at a time. Thus it is not possible to create a printout with a Gantt Chart and a Resource Sheet (Table) or Graphs (Histograms) with Microsoft Project. Oracle Primavera P6 allows two panes to be printed at a time, a Gantt Chart from the top pane and a histogram in the bottom pane. Asta Powerproject allows a Gantt Chart plus multiple Histograms to be printed at one time.

3.4.6   Managing Details Forms

Some **Views** displayed in **Panes** have further options for displaying data. These are titled **Details** forms. The **Details** forms may be selected, when available, by:

- Making the pane active, then
- Right-click in the right-hand side of the screen and clicking the required form, see the example to the left of this menu:

3.5   Status Bar

The **Status Bar**, located at the bottom of the screen, may be formatted by right-clicking on it.

Switch on the Macro Recorder option if you use macros.

 It is recommended that the **Zoom Slider** is not displayed as the use of this function results in very strange timescale scale increments.

3.6 Forms Available from the Ribbon Groups

Some Ribbon groups have a little arrow in the bottom right-hand corner of the Group box. Clicking on the arrow will open up a form:

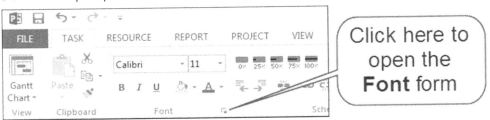

3.7 Right-clicking with the Mouse

It is very important that you become familiar with using the right-click function of the mouse as this is often a quicker way of operating the software than using the menus.

The right-click will normally display a menu, which is often different depending on the displayed View and Active Pane. It is advised that you experiment with each view to become familiar with the menus.

3.8 Finding the Task Bars in the Gantt Chart

When there are no bars displayed it is sometimes difficult to find them.

To find a task bar:

- Select a task that you wish to find the corresponding bar,

- Select the **Task**, **Editing** group, **Scroll to Task** button,

- This will move the Gantt Chart Timescale to display the task bar.

3.9 Setting up the Options

The basic parameters of the software must be configured so it will operate the way you desire. In order for the software to operate and/or calculate the way you want, some of the defaults must be turned on, or off, or changed. These configuration items may be found under **File**, **Options**.

This book assumes you have a default install of Microsoft Project 2013 or 2016, either Standard or Professional versions, and the Options and Global template have not been edited. We will discuss some of the more important Options now. All the Options are discussed in detail in the **OPTIONS** chapter. If you have a non-standard install of Microsoft Project you may need to read the **OPTIONS** chapter to check and understand how your software is set up.

The options in the first workshop should be used by new users and changes as required.

Ensure **ALL PROJECTS ARE CLOSED** before you start setting your options then all changes that you make to the options will be applied to new projects created using the **File**, **New** command which uses your **Global.mpt** template.

3.9.1 General Tab

Select **File**, **Options** to display the **Options** form and select the **General** tab:

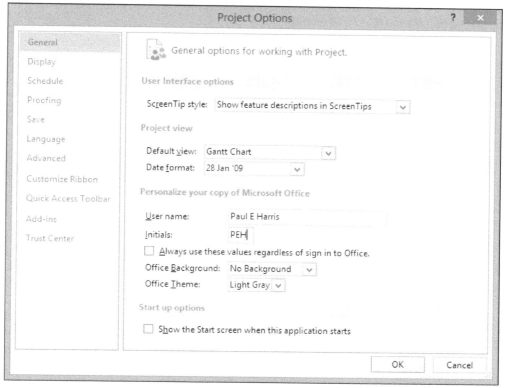

- **Default view:** is the view that is applied to any project when it is opened. It is recommended to set this to **Gantt Chart** so the **Timeline** view is not displayed by default.

- **Date format:** formats the dates and time display style (12 or 24 hour clock) for all projects. The date and time format will be displayed according to a combination of your system default settings and the Microsoft Project Options settings. You may adjust your date and time format under the system **Control Panel**, **Region and Language Options** and the Microsoft Project settings in the Options form, which are covered in the **OPTIONS** chapter para 22.1.2 Project view .

 There is often confusion on international projects between the numerical US date style, mmddyy and the numerical European date style, ddmmyy. For example, in the United States 020719 is read as 07 Feb 19 and in many other countries as 02 Jul 19. You should consider adopting either the ddmmmyy style, **06 Jan '19** or mmmddyy style, **Jan 06 '19** for all your plans to avoid these misunderstandings especially when the project involves parties from within and outside the US on one project.

- **Start up options:** unlike earlier versions of Microsoft Project, a blank project titled **Project 1** is not automatically created, so you will either need to:

  ➢ Create a new project or open an existing project before you may start working, or

  ➢ If you wish the software to start up with the Gantt Chart View with a blank project titled **Project 1** created (as in earlier version of Microsoft Project) then you will need to select **File**, **Options**, **General**, **Start up options** and uncheck the **Show the Start screen when this application starts**: and set the **Default view:** to the **Gant Chart**,

---

3.9.2 Schedule Tab – Scheduling options for this project:

Many of the default Microsoft Project Schedule options make the software operate at a more complex level making the software difficult for all users. It is suggested that the settings discussed below are considered as a starting point for all projects.

Select the **Schedule** tab and scroll down to the **Scheduling options for this project:**

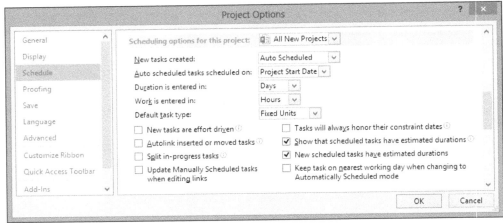

- **Scheduling options for this project**: allows the selection of **All New Projects** or any project that is currently open from the drop-down box.
  - ➢ When **All New Projects** is selected then the defaults selected in this form are applied to any new project when it is created from the **Blank project** option. These options are **NOT** used when a project is created from a template because the template options are adopted for new projects created from a template.
  - ➢ When an open project is selected then the options for that project may be edited.
- **New tasks created:** allows the selection of **Manually Scheduled** or **Auto Scheduled**.
  - ➢ It is normally considered good scheduling to select **Auto Scheduled** so tasks will acknowledge the relationships and constraints.
  - ➢ The **Manually Scheduled** option (new to Microsoft Project 2010) overrides the schedule calculations for tasks marked as **Manually scheduled**, allowing the software to be used like a white board for the selected tasks. This feature is covered in more detail in the **ADDING TASKS** chapter.
- **Auto scheduled tasks scheduled on:** – When the option of **Start on Current Date** is selected new tasks are assigned an Early Start constraint as they are added to the schedule.

 It is not desirable to have tasks assigned constraints as they are created in a Critical Path schedule as this affects how the dates and float are calculated. Therefore **Start On Project Start Date** should always be selected when creating a Critical Path schedule.

- **Duration is entered in:** – This option specifies the format in which durations are entered via the keyboard. If **Day** is selected as the default, then a duration of 2 days is entered as 2 (without the d). If **Hours** is selected as the default, then a 2-day duration should be entered as 2d. Normally **days** are selected here unless your project is a very short duration and you wish to display the durations in hours. The author normally selects **Days**.
- **Default task type:** – This option becomes active when resources are assigned and should be set to either **Fixed Duration** when the user does not want the duration of tasks to change as resources are modified or **Fixed Units** when the user prefers the estimate at completion not to change as task durations are edited. This topic is covered in more detail in the **OPTIONS** and **RESOURCES** chapters. The author often uses **Fixed Units** as this option displays progress on

Split Tasks better than the **Fixed Duration** option. Both options have advantages and disadvantages.

- **New tasks are effort driven** – This option becomes active when resources are assigned to tasks and the author recommends this should be unchecked. The reasons are covered in detail in the **OPTIONS** and **RESOURCES** chapters. If this is left checked and more than one resource is assigned to a task then either the Duration or resource Units per Time Period will change which may become confusing for new users. The author normally unchecks this option.

- <u>A</u>utolink inserted or moved tasks option will result in relationships being changed when tasks are dragged to another position. This should **NOT** normally be checked.

- **S<u>p</u>lit in-progress tasks** will make some tasks split when they actually start before their predecessors. This can get confusing for new users when tasks generate splits. This function also creates a more conservative schedule normally taking longer to complete the tasks. The author suggests that new users turn this option off until they have some experience with the software.

- **Update Manually Scheduled tasks when editing links** makes **Manually Scheduled** tasks acknowledge their relationships when they are assigned a predecessor. A **Manually Scheduled** task will not move in time if the option is checked or not when it does not have a predecessor. The author does not recommend using **Manually Scheduled tasks** so the selected option is not important.

- **Tasks will alwa<u>y</u>s honor their constraint dates** results in the possibility of tasks being scheduled earlier than it is technically possible. This should **NOT** ever be checked.

- **Estimated Durations** – These two options do not affect the calculation of projects. When checked a new task is assigned an estimated duration and has a "**?**" after the duration. Once a duration is manually entered, the "**?**" is removed. The **General** tab on the **Task** form has a check box which establishes when a task has an estimated duration.

  - ➤ **<u>S</u>how that scheduled tasks have estimated durations** – A task with an estimated duration is flagged with a "**?**" after the duration in the data columns when this option is checked.

  - ➤ **New tasks ha<u>v</u>e estimated durations:** – When a task is added, it will have the **Estimated** check box checked in the **Task Information** form.

  Before entering a Duration   and after entering a Duration:

Task Name	Duration
New Task	1day?

Task Name	Duration
New Task	6days

Summary tasks are marked as estimated when one or more **detail tasks** associated with it are marked as Estimated.

- **Keep task on nearest working day when changing to Automatically Scheduled mode** assigns a constraint to a task when it is changed from **Auto scheduled** to **Manually scheduled**. Therefore an **Auto scheduled** task does not move to the **Project Start Date** when converted to **Manually scheduled**. Constraints should be set by the user. It is bad practice to allow the software to set constraints for the user and it is recommended this option is turned off.

3.9.3   Schedule Tab – Advanced Options

These are covered in detail in the **Project Options** chapter but should be set as per **Workshop 1**.

3.10   Mouse Pointers

There are a number of mouse pointers and this table will outline the important ones:

Mouse Pointers when hovering over a table:

Normal mouse pointer used to grab or click

Select cell

Adjusts width of a row

Select row

Select column

Drag one or more rows or columns to move them to a new location

Drag Cell

Displayed when hovering over the Task name and used for indenting and outdenting tasks

Mouse Pointers when hovering over a bar in the Gantt Chart:

Increase or decrease task duration

Select task, before linking or dragging a task

After selecting a task, dragging left or right will move the task and set a constraint

After selecting a task, dragging up or down will add a Finish-to-Start relationship

When dragged right will assign an Actual Start and adjust the % Complete

Other Pointers:

Adjusts the width of tables in forms such as the width of columns in the Task Form

Moves the divider between the Table and Gantt chart

Moves the divider between the Upper and Lower Pane

3.11   Select All Button

The **Select All** button should be used to select all tasks in order to copy them or delete them.

3.12 Short Cut Keys

The normal windows short cut keys such as Ctl+C and Ctl+V or **Copy and Paste**, Ctl+B for **Bold**, Ctl+I for **Italic**, Ctl+V and Ctl+Z for **Undo and Redo**, and Ctl+X for **Cut** operate with Microsoft Project.

There are far too many short cut keystrokes available that most people never use and a list may easily be found by searching help, F1 key, or searching on the internet.

The table below lists the more useful special keystrokes that are unique to Microsoft Project or not well known:

Key Stroke	Command
Alt + Shift + Left Arrow	Outdent Task
Alt + Shift + Right Arrow	Indent Task
Alt +F1	Project Information
Alt +F10	Assign Resource
Insert	Creates a New Task
Ctl+A	Select All
Ctl+F	Find
Ctl+G	Go To Task ID or Date
Ctl+H	Find and Replace
Ctl+N	New Blank Project
Ctl+P	Print
Ctl+S	Save
Clt+F2	Links Selected Tasks
Clt+Shift+F2	Unlinks Selected Tasks
Clt+Shift+F5	Scroll to Task
Shift+F2	View Task Information
F5	Go To Task ID or Date
F7	Spell check
F9	Reschedule

3.13 Help - Tell me what to do

Microsoft Project 2016 introduced a function titled **Tell me what to do**. This function is located in the menu bar and typing a question in the box supplied interactive help:

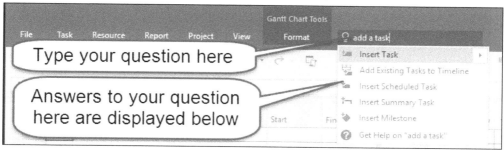

3.14   Workshop 1 – Navigation and Setting Your Project Options

Background

In this workshop you will practice navigating around the screen, set the options to allow durations to be entered in days, ensure that a useful date format is displayed and ensure other options are set so the software operates in a simpler mode than the standard defaults.

Navigation Practice

1.   Click on the Ribbon Toolbar menu at the top of the screen, work your way through the tabs and observe what commands are located on each tab, the toolbars in Project 2013 and 2016 are slightly different:

2.   Right-click on the Ribbon Toolbar and display the Ribbon Toolbar Menu:

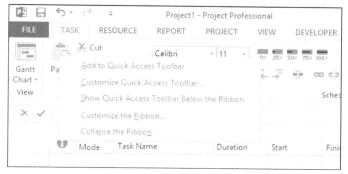

3.   To allow more buttons to be displayed on the Quick Access Toolbar, click on the **Show Quick Access Toolbar Below the Ribbon** to move the Quick Access Toolbar below the Ribbon Toolbar.

4.   From the same menu, click on the **Customize Quick Access Toolbar…** to open the **Project Options** form. This form may also be opened by selecting **File**, **Options**. Now explore the tabs on the left-hand side of the **Project Options** form.

5.   With the **Project Options** form open, click on the **Quick Access Toolbar** tab and either:

   ➢   Download the Microsoft Project Quick Access Toolbar from the www.eh.com.au/ website Software & Downloads page, unzip it by double clicking on the file and dragging it to your Desktop and import the toolbar using **File**, **Options**, **Quick Access Toolbar**, **Import/Export**, or

continued…

> ➢ If you are unable to download the Quick Access Toolbar, then add the following frequently used buttons to the Quick Access Toolbar, if they not already displayed:

6. Click on the **Customize Ribbon** tab in the **File**, **Options** to open the **Project options** form, ensure that **Developer** tab is checked. The **Developer** tab will now be displayed on the **Ribbon** to allow access the **Organizer** form more easily, and then close the form.

7. Right-click on the Ribbon Toolbar and display the Ribbon Toolbar Menu. Click on **Collapse the Ribbon** to hide the Ribbon Toolbar. When you click in the Gantt Chart area the Ribbon will minimize and more work area will be available allowing you to see more tasks. This is useful when you have a small screen.

8. Right-click in the Gantt Chart, select **Show Split** to split the window and this will show the **Task** form in the bottom window.

9. Make the bottom window active by clicking in it.

10. Note the text on the left-hand side of the screen is highlighted, when moving from the top pane to the bottom pane. The active pane has the highlighted text. This may be quite hard to see with some screen colors.

11. Right-click in the bottom pane and select the different menu options to see how the **Task Details** form changes with the different options. Leave this form with the **Predecessors and Successors** option displayed.

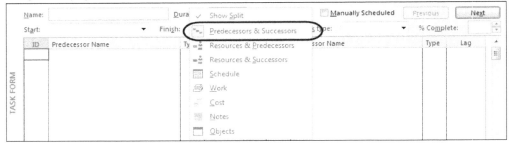

12. Activate the upper pane, by clicking in it.

13. Resize the panes by dragging the Split screen bar.

14. Close the Split screen by double-clicking on the horizontal dividing line.

15. Split the screen by double-clicking on the small bar in the bottom right-hand corner of the screen.

Assignment – Set the Options

1. Close your project by selecting **File**, **Close** and do not save any changes, you should have Microsoft Project open, but no projects in view and a blank screen.
2. Select **File**, **Options** to open the **Project Options** form.
3. Select **General** tab and set the **Default View** to Gantt Chart.
4. Set the **Date format:** to either:

   ➢ "**ddmmmyy**" i.e., 28 Jan '09, or
   ➢ "**mmmddyy**" i.e., Jan 28 '09.

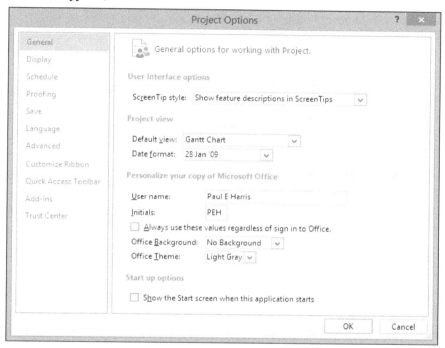

NOTE: The available date format will depend on your system settings, set in the system Control Panel, Region and Language Options. If you wish to show the time in 24 hour format then there is more information in the **OPTIONS** chapter, para 22.1.2 Project view.

5. Enter your name and initials.

6. If you uncheck the **Start up options**, **Show the Start screen when this application starts** then the **Start Screen** will not be displayed when you start Microsoft Project and you will be taken straight to the **Gantt Chart** View.

7. Select the **Display** tab and check that **ALL** check boxes are checked.

8.   Select the **Schedule** tab and set the Schedule Options for your project as per the picture below:

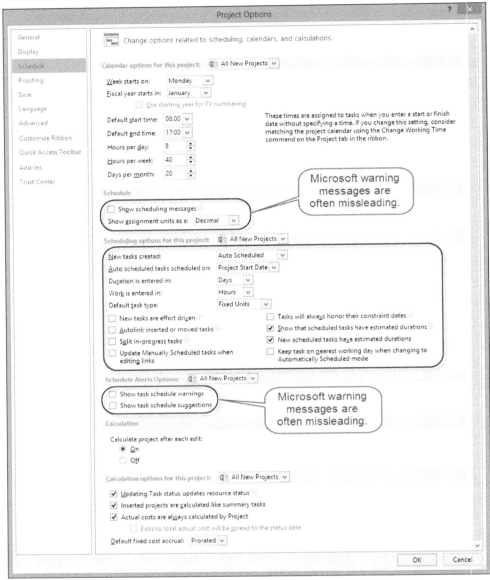

NOTE: The picture above shows the time in 24 hour format, if you wish to show the time in 24 hour format then you will need to change your system setting in Control Panel, Region and Language Options. There is more information in the **OPTIONS** chapter, para 22.1.2 Project view.

9.   More details about the settings in the form above, and how they operate, are available in the **OPTIONS** chapter.

Continued…

10. Select the **Advanced** tab and set options as per the diagram below:

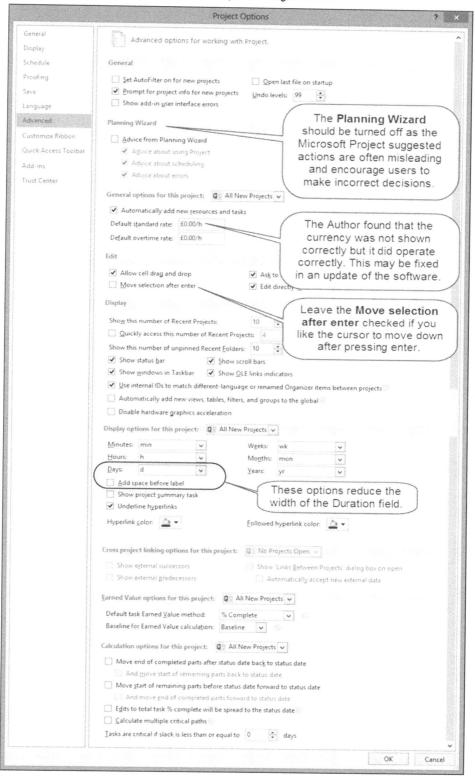

11. Go to the **Trust Center**, **Trust Center Settings…**, **Legacy Formats** and select **Prompt when loading files with legacy or non-default file format**. This option will allow you to open files created in earlier versions of Microsoft Project but will warn you that it is not a Microsoft Project 2010 to 2016 file format.

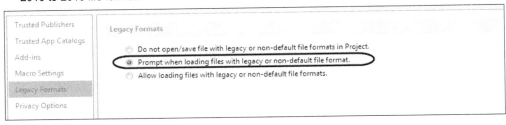

12. Select [OK] twice to close the **Project Options** form.

4 CREATING PROJECTS AND TEMPLATES

4.1 Starting Microsoft Project

Ensure you have Automatic Updates switched on, this is part of Microsoft office, as the product is often improved over time and bugs removed.

There are three principal methods of creating a new project:

- Create a blank project from the **Global.mpt** template, this template is part of your system, or

- Create a project a user template that contains default data and formats. This will require you to set the **Save template** directory in for this function to operate, see next para and this is covered in detail later in this chapter in para 4.5, or

- Open an old project and save it with a new file name.

When opening Microsoft Project, you will NOT be presented with a blank project as in earlier versions, and you will need to open or create a project before you may start working unless you check the **Project Options**, **General**, **Start up options**, then uncheck the **Show the Start screen when this application starts**: check box

4.2 Creating a Blank Project

A blank project may be created from the **New Project** pane, which is displayed by:

- Keying in **Ctrl+N**, or

- Selecting **File**, **New**,

- Then clicking on **Blank Project**

A new project created from your **Global.mpt** will be displayed and the project name is shown at the top center of the screen.

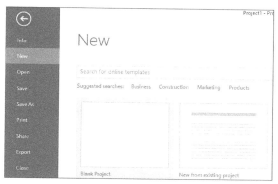

At this point the **Project Start date** is normally set in the **Project Information** form. Select **Project**, **Properties** group, **Project Information** to open this form and the time if not displayed, is set by default from your **Options** usually at 08:00hrs or 08:00pm:

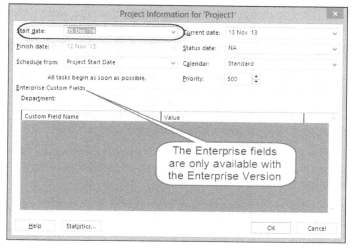

- When **Schedule from:** is set to the **Project Start Date**, which is the usual method of scheduling projects:
  - ➢ Enter the **Start date:** – This is the date before which no task will be scheduled to start.
  - ➢ The **Finish date:** – This is a calculated date and is the date of the completion of all tasks.
- When **Schedule from:** is set to the **Project Finish Date**:
  - ➢ All new tasks are set with a constraint of **As Late As Possible**, and you will not be creating a critical path schedule required by many contracts.
  - ➢ Summary Tasks are set to **As Late As Possible**, and
  - ➢ Therefore, all new tasks are scheduled before the **Project Finish Date** and not after the start date, which is now calculated by the software.

 This option of scheduling from the **Project Finish Date** is not recommended as the schedule calculation gives some results that are difficult to understand because all the tasks are scheduled As Late As Possible. This also does not create a Critical Path schedule which is required in most contracts. This topic is covered in paragraph 24.6.5.

- **Current Date:** – This field defaults to **today's date**; it represents the date today and may be changed at any time. This date has no effect on most calculations and reverts back to the system date each time a schedule is opened. The time is by default set to 8:00am or 08:00hrs.

 The **Current Date** is set to the system date each time a schedule is opened. It is recommended that this date not be used for identifying the **Data Date**. The **Status Date** does not change and this date should always be used for identifying the **Data Date**.

- **Status Date:** – This is an optional field used when updating a project. This topic is covered in the **TRACKING PROGRESS** chapter. After this date has been set, it may be displayed as a gridline. You may remove the date by typing **NA** into the field.
- **Calendar:** – This is the project **Base** calendar that is used to calculate the durations of all tasks unless they have:
  - ➢ A resource with an edited resource calendar, or
  - ➢ A different task calendar assigned.
- **Priority:** – This is the project priority when sharing resources over a number of projects. The high priority is 1000 is and the lowest is 0.
- Click on the [Statistics...] button to open the **Project Statistics** form, which outlines statistical information about the project.

A new blank project copies default values such as the Standard calendar from the **Global.mpt** file. The **Global.mpt** file is a template on your computer used to create blank projects and this may be edited using the **File**, **Info**, **Organizer** utility.

 The default Microsoft Project blank project has a Standard calendar based on 5 days per week without any holidays, which will not suit many projects. To avoid setting up new calendars for each project it is recommended that you either:

  - • The Standard calendar be replaced with a project calendar in the **Global.mpt** that has been edited to represent your local public holidays using the **File**, **Info**, **Organizer** utility, or

  - • Consider using Templates to create new projects which have had the calendar edited to suit your organization's work periods.

4.3   Opening an Existing Project

Another method of creating a new project is to open an existing project, save it with a new name and then modify it. To open an existing project display the **Open** form by selecting:

- **File**, **Open**, or

- **Ctl+O**

Then select the location that you wish to open a file from:

- **Recent Projects**

- **OneDrive**

- **Computer**

- **Add a place**

The picture on the right shows the **Open** form that is displayed when **Computer, Browse** is selected:

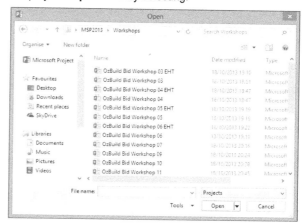

The **New from existing project** form may be used to copy an existing project:

- Select **File**, **New**, **New from existing project**,

- Then select the file you want to open from the **Open** form,

- Then use **File**, **Save As** to save the file under a different name,

  ➢ Enter a new **File Name**,

  ➢ Select in which **Current Folder** you want to save the project, and

  ➢ Click on ⬚ Save ⬚ to save the new project.

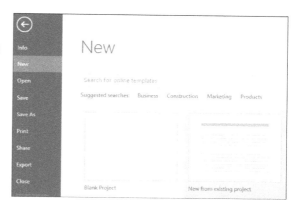

You may now alter the contents of this existing plan to reflect the scope of your new project.

4.4   Re-opening a Project

Selecting **File**, **Open**, **Recent Projects** will show a list of recently opened project files:

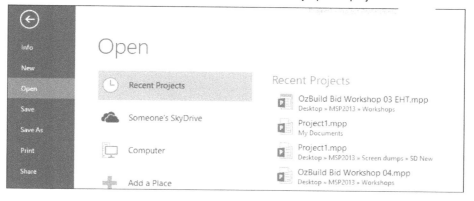

4.5   Creating a Project Template

Project templates allow organizations to create standard project models containing default information applicable to the organization. In particular, a calendar with the local public holidays should be created along with Views with redefined headers and footers to suit your organization. Therefore to save time when you create a new project, you should generate your own templates to suit the different types of projects your organization undertakes.

4.5.1   Setting the Personal Template Directory

To create and use templates for new projects you firstly create the directory to save the templates in:

- Select or create a location on your hard drive or network to save your templates, typically using Explorer,

- Select **File**, **Options**, **Save** and set your **Default personal templates location** here,

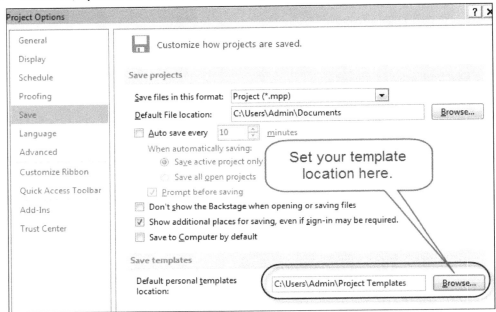

 Company templates could be created and save to a shared drive where everyone could access the latest template.

- At this point in time when you select **File**, **New** there will be an additional option for **Featured** and **Personal Templates**:

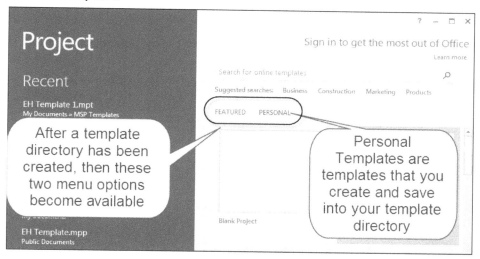

4.5.2   Creating a Personal Template

To create a template:

- Saving a project, with or without tasks, in **Template (*.mpt)** format in your Microsoft Project Template directory.

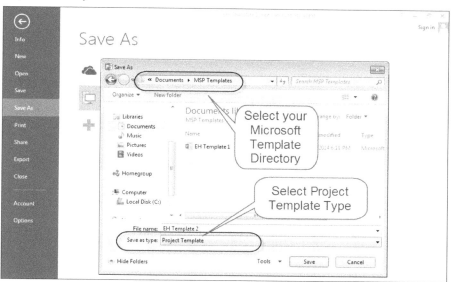

- You will be given some options when saving the template.

- This template will be available when you select **File**, **New**, **My templates**.

- This directory where user templates are saved may be changed by selecting the **File**, **Options**, **Save** tab.

- Templates may be deleted by selecting the template in the **New** (project) form, right-clicking to open a menu and selecting Delete.

4.6   Creating a New Project from a Template

To create a new project from a template:

- Select **File**, **New**:

  > **Featured** are similar to the Microsoft Project 2010 **Office.com Templates**, this will allow you to access Microsoft's templates.

  > **Personal** allows you to open templates on your computer and is covered in the next paragraph.

- Click on the **Personal** and after you have created your own templates these will be available from the **Personal Templates** tab:

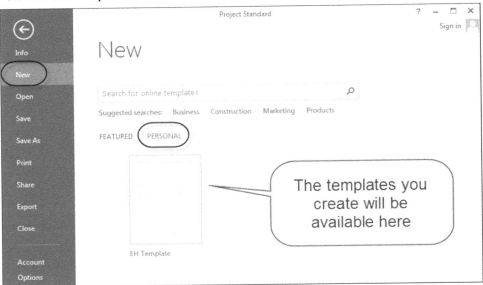

- Select the required template from the form by double-clicking on the template button and then you set the project start date in the next form:

- At this point in earlier versions of Microsoft Project you would normally set the **Project Start date** in the **Project Information** form by selecting **Project**, **Properties** group, **Project Information** to open this form. But in Microsoft Project 2013 and 2016 this date has now been set.

4.7   Saving Additional Project Information

Often additional information about a project is required to be saved with the project such as location, client and type of project. This data may be saved in the **File**, **Info**, **Project Information**, **Advanced Properties** form and these fields may be inserted into printouts and reports:

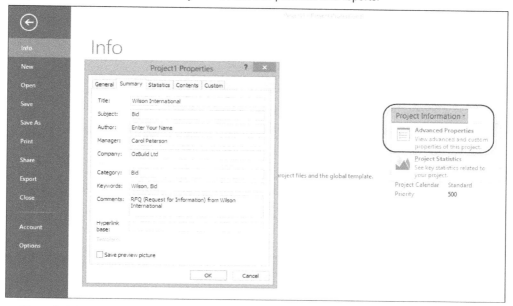

- **Hyperlink base:** This allows you to enter the path to a file or web page.
- **Save preview picture**. This saves a thumbnail sketch, which may be seen when viewing files in Windows Explorer.

4.8   Using the Alt Key and Keystrokes to Access Commands

Microsoft Project has a function that displays the keystroke commands when the **Alt** key is pressed. The picture below shows the keystroke commands after pressing and holding down the **Alt** key; then, for example, pressing the **F** key will open the **File** menu:

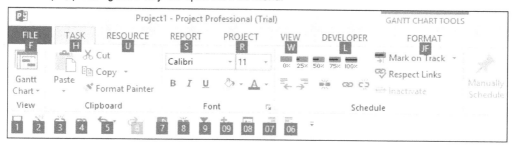

This is useful if your mouse has stopped working and enables you to save a file before rectifying the problem.

4.9   Saving a Project

When you save a project you should consider using a file naming convention that simply allows the identification of the project name or number, the **Status Date** of the file, the type of schedule and version. Many schedulers use the international standard date format of yyyymmdd as the first 8 characters of the file name as this format sorts the files in date order in the directory they have been saved into.

To save a project file use the **File**, **Save** and **File**, **Save As** commands.

A suggested file format would be:

yyyymmdd_"Project Number"_"Revision Number"_"Schedule Type"

yyyymmdd would be in numbers which allows the files to be sorted in date order in Explorer, thus 20120919 would be 9th September 2019.

4.10   Closing Microsoft Project

If you have made changes to a project you will be asked on closing if you wish to save the project.

To close a project:

- Select **File**, **Close**,

- Click on the **Close Window** button and when the last window of a project is being closed you will be asked if you wish to save any changes

- Click on the **Close** button to close all projects and close Microsoft Project.

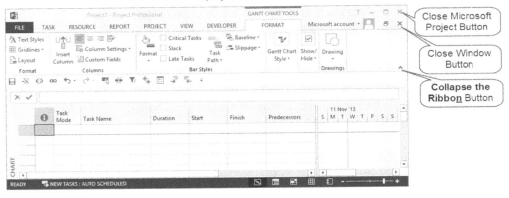

4.11 Workshop 2 - Creating a Project

Background

You are an employee of OzBuild Ltd and are responsible for planning the bid preparation required to ensure that a response to an RFQ (Request For Quote) from Wilson International is submitted on time. Your company has completed the Startup Phase of the project, the Bid Strategy has been developed, and approval to bid for this project has been given. You have been requested to plan the project Initiation Phase, where the Bid will be produced and submitted to the customer. You have been advised that the RFQ will not be available until the 03 December 2018. The Bid will comprise the following deliverables/products:

- Technical Specification

- Delivery Plan

- Bid Document.

These workshops will take you through the process of creating a schedule for the development of the Bid, which will be submitted in response to the RFQ.

 A project template in mpp format has been loaded on the Eastwood Harris web site at www.eh.com.au Software Downloads page that has a number of the issues with Microsoft Project defaults and other setting resolved. You should download this file, open it and save it as a template and use this file instead of the **Blank Project** option as it has some formatting issues resolved.

Assignment

1. Create or select a directory to save your Microsoft Project templates in Windows Explorer.
2. Select **File**, **Options** to open the **Project Options** form.
3. Select the **Save** tab and set the directory that your Microsoft Project Templates will be saved in:

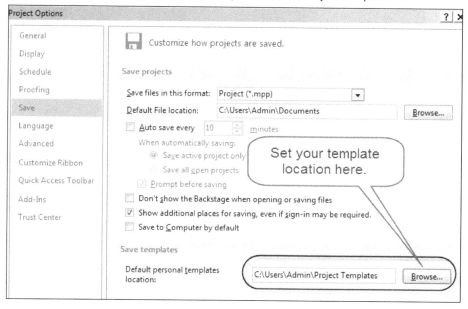

4. Download the Microsoft Project 2013 and 2016 template from www.eh.com.au - Software Downloads page.

5. Unzip it by double clicking on the zip file and dragging the file to your desktop.

6. Open the file with Microsoft Project.

7. Select **File**, **Save As**, **Computer** and **Browse**, then, select from the drop down box to save as type **Microsoft Project template (*.mpt)**. Microsoft will default to the directory you have selected as a Template directory, so save the file in the Template Directory.

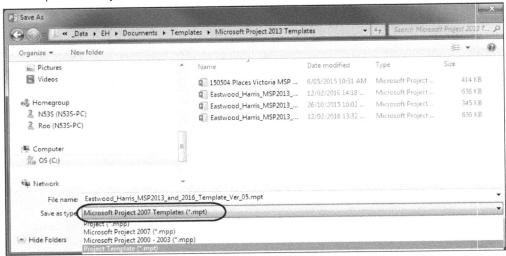

8. Select the defaults at the **Save As Template** form and click **Save**.

9. Close the Eastwood Harris template file and any other projects that may be open.

10. Then create a new project using this template, with the command **File**, **New, Personal** and select the Eastwood Harris template.

11. If you click once on the template, you will get the screen shown below, where you are able to set the **Project Start Date.** Enter 03 Dec 2018, then click on **Create**:

12. If you double clicked on the template, the project will have opened in the Gantt Chart view and you will need to set the **Start date:** to **Mon 03 Dec 18** using the **Project, Properties** group, **Project Information** form, but do not edit the **Current date:** Press the [OK] button to save the input data.

13. Now check again the project **Start date in** the **Project**, **Properties** group, **Project Information** form:

NOTE: The date format will be displayed according to a combination of the Microsoft Project Options settings and your computer system default settings. You may adjust your date format so you are able to see the day month and year using the **File**, **Options**, **General** form. The order of the day, month and year is set in the system Control Panel, Region and Language Options.

16. **Save** your project as **OzBuild** in a location of your choice, such as the desktop.

17. Add the following project information in the **File**, **Info**, **Project Information** (Drop down box on the top right hand side of the screen), **Advanced Properties** form.

18. **Save** your project as **OzBuild** again.

NOTE: Completed workshops and PowerPoint Instructors slide presentations may be downloaded from the Eastwood Harris web site at www.eh.com.au. This book forms the basis of a Project Management Institute (PMI) registered course and training organizations who wish to award PMI Professional Development Units (PDUs) may use this book and PowerPoint slide show as a licensed training course with the author's permission. Please contact the author Paul E Harris email: harrispe@eh.com.au.

This publication is only sold as a bound book, no parts may be reproduced by any means, e.g. electronic, video or print.

© *Eastwood Harris Pty Ltd*   **54**

5   DEFINING CALENDARS

5.1   Understanding Calendars

The finish date (and time) of a task is calculated from the start date (and time) plus the duration over the calendar associated with the task. Therefore, a five-day duration task that starts at the start of the work day on a Wednesday, and is associated with a five-day workweek calendar (with Saturday and Sunday as non-work days) will finish at the end of the workday on the following Tuesday.

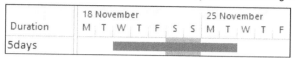

Microsoft Project is supplied with three calendars, which are termed **Base Calendar**s:

- **Standard** – This calendar is 5 days per week, 8 hours per day.
- **24 Hour** – This calendar is 7 days per week and 24 hours per day.
- **Night shift** – This calendar is 6 days per week and 8 hours per day during the night.

You may create new or edit existing Base Calendars to reflect your project requirements, such as adding holidays or additional work days or adjusting work times. For example, some tasks may have a 5-day per week calendar and some may have a 7-day per week calendar.

A **Base Calendar** is assigned to each project. Microsoft Project uses the term **Project Calendar** to describe this calendar. By default all un-resourced tasks use the **Project Calendar** to calculate the end date of the task. Any un-resourced task may be independently assigned a different calendar and the end date will be calculated using the assigned calendar.

This chapter will cover some of the following topics, those with an * are covered in the **MORE ADVANCED SCHEDULING** chapter.

Topic	Menu Command
• Assigning a base calendar to a project	**Project**, **Properties** group, **Project Information**
• Editing a calendar's working days	**Project**, **Properties** group, **Change Working Time**
• Creating a new calendar	**Project**, **Properties** group, **Change Working Time**, **Create New Calendar…**
• *Renaming an existing calendar	**Developer**, **Manage** group, **Organizer**, **Calendars** tab, **Rename…**
• *Deleting a calendar	**Developer**, **Manage** group, **Organizer**, **Calendars** tab, **Delete**
• *Copying a calendar to Global.mpt for use in future projects	**Developer**, **Manage** group, **Organizer**, **Calendars** tab, **Copy >>**
• *Copying calendars between projects	**Developer**, **Manage** group, **Organizer**, **Calendars** tab, **Copy >>**

Microsoft Project 2007 introduced significant changes in the way calendars are created and edited and which now allows for each non-work day or period to have a name or note explaining the purpose of the non-default work period. These non-work days or periods could be due to holidays or changes to working hours for a planned shutdown or system upgrade requiring people to work different or longer hours.

5.2 Editing Calendars

5.2.1 Editing Working Days to Create Holidays

To edit a calendar, select **Project**, **Properties** group, **Change Working Time** to open the **Change Working Time** form:

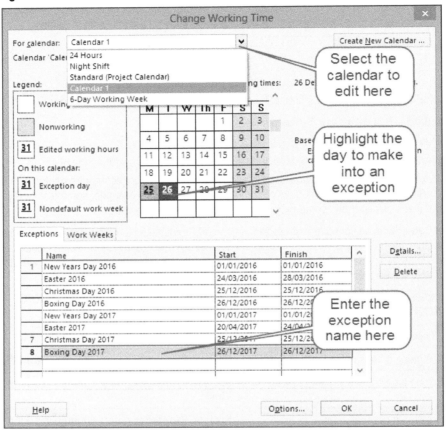

- The **Project Calendar** is also identified in the **Change Working Time** form as the calendar with **(Project Calendar)** written after the calendar name.
- Select the calendar to be edited from the **For calendar:** drop-down list at the top of the form,
- Click on the **Exceptions** tab,
- Highlight the days to be made nonworking days in the calendar,
- To make **Work Days** into **Nonworking Days**, add the following information in the first blank line in the lower half of the form:
  - ➢ Name of the Nonworking Day,
  - ➢ The Start Date, and
  - ➢ The Finish Date

NOTE: The Start and Finish Dates of a one-day holiday are identical.

- When the holiday is recurring, such as the annual Christmas Day and New Year's Day, these days may be made recurring by clicking on the [Details...] button to open the calendar **Details:**

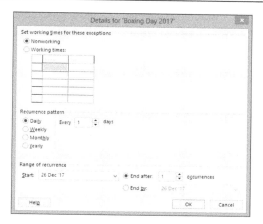

  > In the **Recurrence pattern** set the occurrence of the holiday to use the Daily, Weekly, Monthly or Yearly options. The options are displayed below:

Daily

Weekly

Monthly

Yearly

  > The **Range of recurrence** options allows the specification of the number of times the holiday occurs. The picture below shows Christmas Day is scheduled for the next 20 years.

Daily

Only single day nonworking periods may be made repeating.

5.2.2 Editing a Calendar to Create an Exception

When it is desired to make a non-work day into a workday, say to make a Saturday a workday in a 5-day per week calendar:

- Open the **Change Working Time** form,

- Select the Calendar in the **For calendar:** drop down box,

- Select the day to be a non-workday,

- Enter the workday description a blank line in the **Exception** tab,

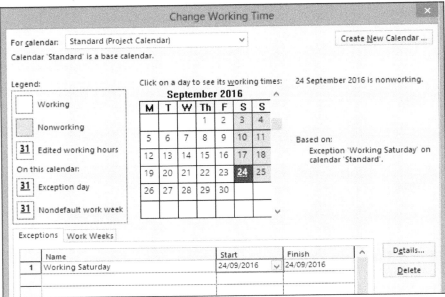

- Click on the [D**e**tails...] button to open the **Details** form,

- Click on the **Working times:** radio button and close the form:

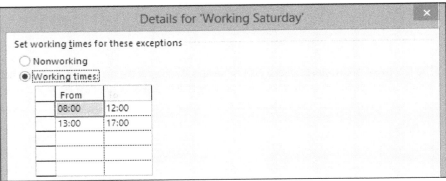

5.3   Creating a New Calendar

Select **Project**, **Properties** group, **Change Working Time** button and click on the
Create New Calendar ... to open the **Create New Base Calendar** form:

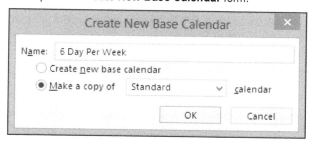

- To create a new calendar which is a copy of the Standard calendar, just type in the new calendar name in the **Name:** box and click on the OK button to create it, or

- To copy an existing calendar, click on the **Make a copy of** radio button and select the calendar you want to copy from the drop-down box and type in the new calendar name.

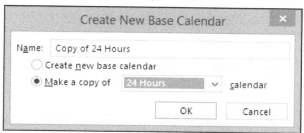

 To save time when multiple calendars have the same nonworking periods, such as religious holidays, one calendar should be created and all of its nonworking periods entered. Then copy this calendar to create new calendars showing the same nonworking days.

5.4   Assigning a Calendar to a Project

In all versions of Microsoft Project a new blank project is assigned the **Standard** calendar as the **Project Calendar** when the project file is created. The **Project Calendar** is changed using the **Project Information** form by:

- Selecting **Project**, **Properties** group, **Project Information**, and

- Selecting the alternative calendar from the **Calendar:** drop-down box:

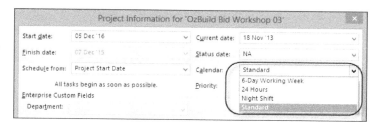

 When a **Project Calendar** is changed or edited, the end date of all tasks assigned with the **Project Calendar** will be recalculated based on the new calendar. A task is assigned the **Project Calendar** when the **Task Calendar** field displays **None**. This may make a considerable difference to your project schedule dates. All new resources are assigned the **Project Base Calendar** when they are created.

© **Eastwood Harris Pty Ltd**

5.5 Calculation of Tasks in Days

Microsoft Project effectively calculates in hours and the value of the duration in days is calculated using the parameter entered in the **Hours per day:** field in the **File**, **Options**, **Schedule** tab, **Calendar options for this project:** section. It is **VERY IMPORTANT** that all users understand that the durations in days are calculated using **ONLY** the one parameter and this parameter is used with **EVERY** calendar. This makes Microsoft Project difficult to use when calendars with a different number of hours per day are being used.

For example, the **Options** form below shows the **Hours per day:** as "8". Therefore, when tasks are:

- Assigned an 8-hours per day calendar then the durations in days will be correct, and
- Assigned a 24-hours per day calendar the durations of tasks in days will be misleading.

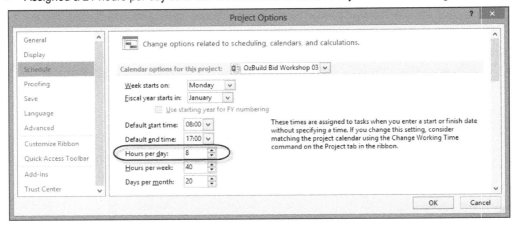

The picture below shows:

- Task 1 has the correct duration in days.
- Task 2 shows a duration that is clearly misleading.
- Task 4 and 5 display the duration in hours and are not misleading when the calendar column is also displayed.

	Task Calendar	Duration	Start	Finish	Mon 01	Tue 02	Wed 03	Thu 04	Fri 05
1	8 Hours per Day	5 days	Mon 08:00	Fri 17:00					
2	24 Hours per Day	5 days	Mon 08:00	Wed 00:00					
3									
4	8 Hours per Day	40 hrs	Mon 08:00	Fri 17:00					
5	24 Hours per Day	40 hrs	Mon 08:00	Wed 00:00					

The **Calendar options for this project** are covered in more detail in the **OPTIONS** chapter.

 It is **STRONGLY RECOMMENDED** that inexperienced users avoid changing calendar start and finish times or assigning calendars with different numbers of hours per day whenever possible.

To display the duration in hours:

- The **Duration is entered in:** field in the **File**, **Options**, **Schedule** tab, **Schedule** heading should be set to **Hours**, and then

- Existing durations in days are changed to hours by overtyping the original durations.

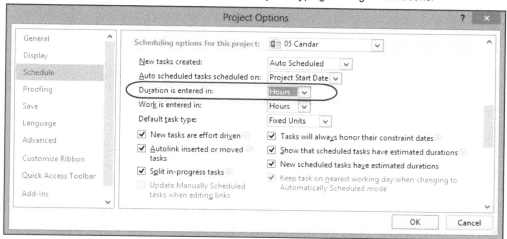

5.6   Effect on 2007 Calendars When Saving to 2000 – 2003

When a project file is saved from Microsoft Project 2007 or later format to 2000 – 2003 formats all the calendar notes are lost and each repeating nonworking day is converted onto individual nonworking days without the note indicating the type of holiday it represents. It is recommended that you do not save to the 2000 – 2003 formats and then reopen the project. The pictures below show the effect before and after saving to 2000 – 2003 formats and then reopening with a later version of Microsoft Project.

Before saving to 2000 – 2003 format   After saving to 2000 – 2003 format

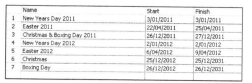

	Name	Start	Finish
1	New Years Day 2011	3/01/2011	3/01/2011
2	Easter 2011	22/04/2011	25/04/2011
3	Christmas & Boxing Day 2011	26/12/2011	27/12/2011
4	New Years Day 2012	2/01/2012	2/01/2012
5	Easter 2012	6/04/2012	9/04/2012
6	Christmas	25/12/2012	25/12/2031
7	Boxing Day	26/12/2012	26/12/2031

	Name	Start	Finish
1	[Unnamed]	3/01/2011	3/01/2011
2	[Unnamed]	22/04/2011	25/04/2011
3	[Unnamed]	26/12/2011	27/12/2011
4	[Unnamed]	2/01/2012	2/01/2012
5	[Unnamed]	6/04/2012	9/04/2012
6	[Unnamed]	25/12/2012	25/12/2012
7	[Unnamed]	26/12/2012	26/12/2012
8	[Unnamed]	25/12/2013	25/12/2013

5.7   Selecting Dates

With the introduction of Microsoft Project 2007 there was a slight loss of functionality in selecting dates by the removal of the drop-down box that allowed the selection of months and the ability to scroll by year.

A calendar form is displayed by clicking on a date cell with the mouse pointer:

- The month may be scrolled forward or backward by clicking on the blue arrows,

- A date is selected by clicking on it, and

- When a date is selected an Early Start Constraint will be set.

5.8   Workshop 3 - Maintaining the Calendars

Background

The normal working week at OzBuild Ltd is Monday to Friday, 8 hours per day, excluding Public Holidays. The installation staff works Monday to Saturday, 8 hours per day. The company observes the following holidays:

	2016	2017	2018	2019
New Year's Day	1 January	2 January*	1 January	1 January
Easter	25 - 28 March	14 - 17 April	30 March - 2 April	19 - 22 April
Christmas Day	27 December*	25 December	25 December	25 December
Boxing Day	26 December	26 December	26 December	26December

* These holidays occur on a weekend and the dates in the table above have been moved to the next weekday.

NOTE: Boxing Day, the day after Christmas, is a holiday celebrated in many countries.

Assignment

1. Edit the **Standard Calendar** to ensure that only the holidays above in 2018 and 2019 are present by selecting **Project**, **Properties** group, **Change Working Time**, see pictures below.
2. Make the following annual holidays repeating, say for 10 years when you set the holidays in 2018, by clicking on the [Details...] button after adding the first holiday:
    ➢ New Year's Day 1 January
    ➢ Christmas Day 25 December
    ➢ Boxing Day 26 December
    NOTE: If the holiday falls on a weekend then the next weekday will have to be manually assigned as a nonworking day.
3. Easter 2018 may be a bit tricky to enter as you are unable to drag days across two months in the Calendar form, so you may wish to enter 30 March 2108 as a single day holiday and then edit the Finish date.
4. Exit the **Calendar** form to save the calendar edits.
5. Create a new calendar titled **6 Day Week** for the 6-day week by:
    ➢ Copying the **Standard (Project Calendar)** using the [Create New Calendar ...] button, then,
    ➢ Select the **Work Weeks** tab, click on the [Details...] tab and make Saturdays work days with the same working hours as the other working days.
6. Save your **OzBuild Bid** project.

Continued......

Answers to Workshop 3

In the edited **Standard (Project Calendar)** of the **Change Working Time** form below, the dates are displayed in the ROW (Rest of World) date format of dd/mm/yyyy. Computers configured with the US date format will see the dates in the mm/dd/yyyy format.

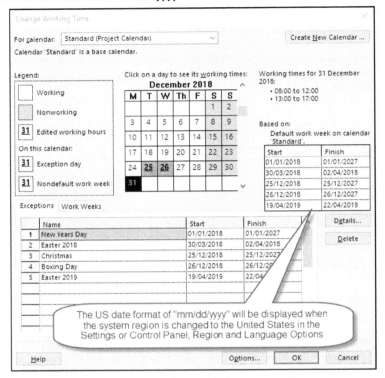

An example of setting Christmas Day as recurring and Saturday as a Workday:

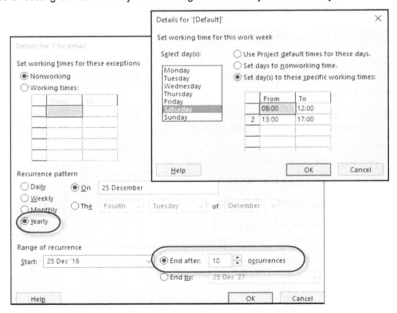

6   ADDING TASKS

6.1   Understanding Tasks

Activities should be well-defined, measurable pieces of work with a measurable outcome. Activity descriptions containing only nouns such as "Bid Document" have confusing meanings. Does this mean read, write, review, submit or all of these? Adequate activity descriptions always have a verb-noun structure to them. A more appropriate activity description would be "Write Bid Document" or "Review and Submit Bid Document."

Adequate task descriptions always have a verb-noun structure to them. A more appropriate task description would be "Specification Approved" or "Specification Issued." The limit for task names is 254 characters, but try to keep task descriptions meaningful, yet short and concise so they are easier to print.

It is considered good practice to first develop a WBS and then develop the tasks required to create the deliverables identified in the WBS.

The creation and sequencing of Detail and Summary tasks are discussed in the following chapters:

- Creating **Detail tasks** in this chapter,
- Creating **Summary** tasks, which may represent the WBS Nodes, in the **ORGANIZING TASKS USING OUTLINING** chapter, and
- Adding the logic in the **ADDING THE DEPENDENCIES** chapter.

This chapter will cover the following topics:

Topic	Menu Command
• **Manually Scheduled or Auto Scheduled**	• **File**, **Options**, **Schedule** tab, **Scheduling options for this project:** section, **New tasks created::**

• **Adding New Tasks**	• Select a line in the schedule and strike the **Ins** (**Insert Key**), or
	• Click on a blank line and type in a cell, or
	• Select **Task**, **Insert** group, **Task** button and there are three options for creating a task,
	
	    Task
	    Recurring Task...
	    Blank Row
	    Import Outlook Tasks...

• **Reordering** tasks	• Select and drag the task(s) or cut and paste in the required location, or • Select **View**, **Data** group, **Sort** to open a menu.
• **Copying** tasks in Microsoft Project	Select the tasks and copy and paste to the required location.
• **Copying** tasks from other programs	Display the required columns and paste the data.
• **Elapsed** duration tasks	Type "e" after the duration and before the Units, e.g. 5 edays.

Topic	Menu Command
• **Milestones**	Assign a zero duration, or Check the **Mark task as milestone** box on the **Advanced** tab of the **Task Information** form.
• **Inactive Task**	A task is marked **Inactive** by checking the option in the **Task Information** form **General** tab.
• **Task Information** form	Double-click anywhere on the task line, or Highlight the task line click on the ⬚ button on the **Task**, **Properties** group **Task Information** button.
• Assigning **Calendars** to tasks	• Set a task calendar in the **Advanced** tab in the **Task Information** form, or • Display the **Task Calendar** column.

6.2   Adding New Tasks

Manually Scheduled tasks will not be covered in detail in this book, we will concentrate on the aspects required to produce a **Critical Path** schedule assuming all tasks are Auto Scheduled.

6.2.1   Adding a Task Under an Existing Task

The first and easiest is to:

- Click on the first blank line under the title **Task Name** and type the task description.

- The duration of the new task is assigned a default duration of 1 day and may have a "**?**" after the duration to indicate it has been assigned an **Estimated Duration**. Overtyping the default duration with the required duration task will remove the "**?**".

- The option of assigning and/or displaying a new task with an **Estimated Duration** is controlled in the **File**, **Options**, **Schedule** tab, **Scheduling options for this project:** section, by checking or un-checking **New scheduled tasks have estimated durations**.

 Summary tasks are marked as estimated when one or more **Detailed tasks** associated with it are marked as Estimated.

- Click into the second row and enter a second Task Name. A sequential **Task Number** starting from "**1**" will be created in the column to the left of the **Task Name**.

The task **Start** and **Finish** dates will be calculated from the **Project Start Date** and the task duration. This information is displayed in the **Start** and **Finish** columns.

6.2.2 Inserting a Task Between Existing Tasks

The second method of adding a task, which is used for inserting a task between two other tasks, is to highlight a task where you want to insert a new task. The highlighted task will be moved down one line after the new task is inserted. Inserting may be achieved by:

- Press the **Insert Key** on the keyboard, or

- Highlight the entire task row by clicking on the Task ID or a cell in a task, then right-click and select **Insert task** from the sub-menu.

Then enter the Task Name, Duration and other scheduling information as required.

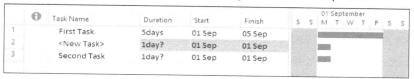

 Do not type in a Start or Finish date into the **Start** or **Finish** column unless you want to set a constraint, which may override the logic.

6.3 Understanding Change Highlight

The Inserted Task in the picture above has the Start and Finish dates highlighted. This is due to the new Microsoft Project 2007 feature titled **Change Highlighting** that highlights any changed dates and durations as a result of an edit, addition or deletion of another task. This function highlights all date and duration fields that have been changed as a result of a change in a Task's dates or duration.

- Remove the highlighting produced by the last change by:
  - ➢ Pressing the **F9** key which will also recalculate the project, or
  - ➢ Saving the project, or
  - ➢ Entering a value into a changed cell twice.

- To change the color of the highlighting select **Format**, **Format** group, **Text Styles** and select **Changed Cells** from the **Item to change** list.

- You should add the **Display Change Highlighting** button (there is no **Change Highlighting** button) to the **Ribbon** (if it does not exit) or add one to the **Quick Access Toolbar** (if it does not exit) to hide/display the highlighting.

This is a useful feature as it highlights changes in all related date and duration fields resulting from a change made to a single task.

6.4 Copying and Pasting Tasks

Tasks may also be copied from another project or copied from within the same project using the normal Windows commands such as **Right-click Copy** and **Paste** or **Ctrl+C** and **Ctrl+V**. You may also copy one or more adjacent tasks by:

- Using the **Ctrl Key** to select rows or dragging and selecting a group of adjacent tasks:

- Ensure you select the whole task or tasks, not just a cell using the Task ID,

- Move the mouse to the new location you want to insert them,

- This will create a copy of the tasks including all assignments like relationships and resources.

 Copy and pasting tasks copies all task data such as calendars, resources and relationships amongst the copied tasks. Unfortunately in Microsoft Project 2013 and 2016, unlike earlier versions, also pastes External Relationships, which are normally not required, and will need to be carefully deleted.

6.5 Copying Tasks from Other Programs

Task data may be copied to and from, or updated from other programs such as Excel, by cutting and pasting. The columns and rows in your spreadsheet will need to be formatted in the same way as in your schedule before they may be pasted into your schedule.

It is recommended that you first display the data columns that you want to import or update in your schedule. If you have no tasks then create a dummy task and data, then copy and paste this dummy data into your spreadsheet, update the data and paste it back into the schedule. Unlike earlier versions of Microsoft Project, the column headers are brought across from Microsoft Project with the option of pasting or not pasting the source formatting.

It is often best in Microsoft Project to make all Task Names unique, so for example when you have a building with many floors and trades each Task Name should include the trade and floor.

- This makes it easier to understand the schedule when a filter has been applied and to find predecessors and successors in a large schedule.

- In Microsoft Project 2013 and 2016 this is even more important because the tasks are listed in the predecessor and successor lists in alphabetical order and not the order in the schedule. So if you have 20 Floors with a concrete activity titled "Concrete", then these activities will be listed all together in a list and it will be difficult to know which floor each concrete task belongs too.

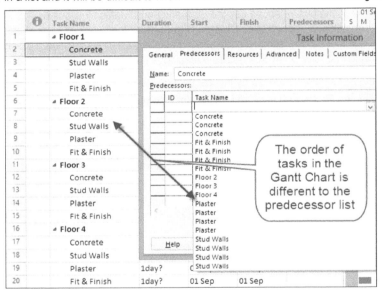

These descriptions may be created in a spreadsheet by using the **Concatenate** function, the picture below demonstrates how. Text may be added by including it in double quotation marks:

	A	B	C
1	Floor 1	Concrete Works	=A1&" - "&B1
2	Floor 1	Frame Walls	Floor 1 - Frame Walls
3	Floor 1	Hang Wall Sheeting	Floor 1 - Hang Wall Sheeting
4	Floor 2	Concrete Works	Floor 2 - Concrete Works
5	Floor 2	Frame Walls	Floor 2 - Frame Walls
6	Floor 2	Hang Wall Sheeting	Floor 2 - Hang Wall Sheeting
7	Floor 3	Concrete Works	Floor 3 - Concrete Works
8	Floor 3	Frame Walls	Floor 3 - Frame Walls
9	Floor 3	Hang Wall Sheeting	Floor 3 - Hang Wall Sheeting

The concatenate function is useful on larger projects where the same tasks will be repeated multiple times. As you can see below the tasks are listed in order in the predecessor list.

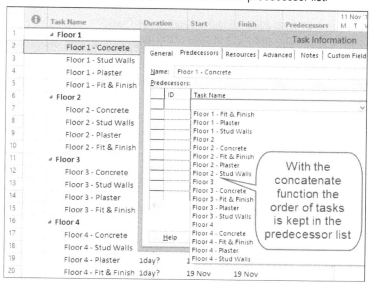

 When you copy and paste dates into the schedule, you may find that activities are assigned constraints, which you may not desire. It is recommended that you display the **Indicators** column, which will show an icon if a constraint has been applied. There is no warning message that these constraints have been set, except a Graphical Indicator in the **Information** column will warn you that a constraint has been set.

6.6   Milestones

A Milestone normally has a zero duration and is used to mark the start or finish of a major event. A Milestone is a Start Milestone when it has no predecessors and is scheduled at the start of a work day, and a Finish Milestone when it has no predecessors and is scheduled at the end of a work day. Other products allow the user to choose if a Milestone is a **Start** or **Finish Milestone** but Microsoft Project does allow this option.

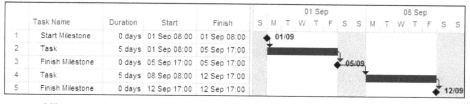

To create a Milestone either:

- Assign a task a zero duration, or

- Click on the **Task**, **Insert** group, **Insert Milestone** button ⌊ 🏷 Milestone ⌋.

Milestones are normally displayed in the bar chart by a ◆.

6.7   Reordering Tasks by Dragging

You may move one or more tasks up or down the schedule by:

- Highlighting any one or more adjacent tasks by using the Task ID column; this will ensure you have selected the whole task and not just cells of the task,

- Moving the cursor to the top line of the selected tasks until it changes to an ⊹ button,

- Then left-clicking and holding down the mouse to drag the row up or down. A gray line will indicate where the tasks will be inserted:

The tasks will be renumbered when in their new location.

6.8   Sorting Tasks

Sorting tasks using fields such as Task name will not renumber them and the sort order may be restored by sorting on the Task ID.

To sort a schedule, select **View**, **Data** group, **Sort** [A/Z↓] to open a menu and select an option.

 In earlier versions of Microsoft Project sorting by some fields, such as Task Name, was ignoring the Outline Level. Thus the project became jumbled up with Detailed Tasks no longer under their assigned Summary Task. This feature has been improved in Microsoft Project 2010 and Detail Tasks are now sorted within each Summary Task but the Summary Tasks are also sorted which is not so useful.

6.9   Task Information Form

The Task Information form may be used to change the attributes of one or more task at one time provided all the tasks are selected before the form is opened.

The Task Information form may be opened in many ways including:

- Double-clicking on a **Detailed Task Line Number**, or **Detailed Task Name**,

- Double-clicking on a **Summary Task Name;** be aware that double clicking on a Summary Task Line Number will only hide or display the Detailed Tasks associated with the Summary Task,

- Clicking on the **Task**, **Properties** group ▭ **Task Information** button when the task is highlighted, or

- Right-clicking in the columns area and selecting **Task Information**, or

- Shift+F2.

You are able to make changes to a number of task parameters, such as the Task calendar, relationships, constraints and resources from this form. Once the form is open, it is not possible to move to another task without closing the form.

There are several tabs on the form:

- **General** – This form contains the basic information about a task:

  ➢ The **Estimated** check box is used to indicate the task has an **Estimated Duration**. The task duration will be displayed with a "**?**" which disappears when a duration is entered against a task.

  ➢ **Priority** is used in resource leveling. The highest is 1000 and the lowest is 0.

  ➢ **Inactive** – This function is only available in the Professional Versions and when a task is marked **Inactive** by checking the option in the **Task Information** form **General** tab then the task is not considered in the scheduling process. Relationships to and from the task are ignored, however as opposed to Microsoft Project 2010 the logic is retained to the predecessor task. Resource Hours and Costs are not included in the =summary tasks rolled up values. The Baseline Values are maintained and this function could be used when some work has been removed from scope and a record of the old scope is required to be maintained:

Below you will see how Microsoft Project 2010 does not retain the logic of the preceeding task.

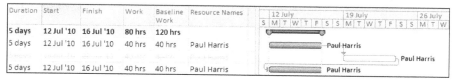

  ➢ **Manually Scheduled** and **Auto Scheduled** are covered in paragraph 24.3.1.

  ➢ The **Bar** options are covered in the **Formatting Bars** section.

- **Predecessors** – This is where the task's predecessors are displayed. This is covered in the **ADDING THE DEPENDENCIES** chapter.

- **Resources** – This is where resources may be created, assigned to tasks and assignment information displayed. This topic is covered in the **RESOURCES** chapters.

- **Advanced** – Options are covered in the **CONSTRAINTS** chapter, paragraph 11.1.1.

- **Notes** – This is where notes about a task may be recorded. This field is handy to record assumptions and is no longer constrained to 256 characters as in earlier versions.

- **Custom Fields** – Any **Custom Fields** that have been customized (by renaming the field) will be displayed in this tab. These are existing fields that may be renamed and customized by selecting **Project**, **Properties** group, **Custom Field**. These fields may hold different types of task information such as text, times, values, etc. There are options to define how these fields are summarized at the summary task level and may be assigned formulae to calculate their values.

- When multiple tasks are selected then the task form may be opened by right-clicking. Some of the attributes of all the tasks may be edited at the same time.

6.10 Elapsed Durations

A task may be assigned an **Elapsed** duration. The task will ignore all calendars and the task will take place 24 hours a day and 7 days per week. A 24-hour, 7-day per week calendar does not need to be created for these tasks. This is useful for tasks such as curing concrete but the Total Float will calculate three times longer than a task on an 8-hour a day calendar and this may be misleading.

To assign an elapsed duration type an "**e**" between the duration and units to enter an elapsed duration. The example below shows the difference between a 7-**Elapsed Day** task and a 7-day task on a **Standard** (5-day per week) calendar.

6.11 Indicators Column

The **Indicators** column, which has an 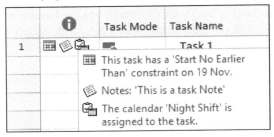 icon in the column header, will display icons in the column when a task contains a non-default setting such as a **Note**, **Constraint**, or a **Task** calendar. Placing the mouse over the icon will display information about the task:

6.12 Assigning Calendars to Tasks

Tasks are by default calculated on the **Project Calendar** and the **Task Calendar** is displayed as **None**. Tasks often require a different calendar from the **Project Calendar**. This would happen when some specific tasks are required to work 7 days per week and the remainder only 5 days per week. In this case the Project Calendar would be set at 5 days per week and the tasks required to work 7 days per week would be assigned a 7 day per week calendar. A Task Calendar may be assigned by using the **Task Information** form or displaying the **Task Calendar** column.

6.12.1 Assigning a Calendar Using the Task Information Form

To assign a calendar using the **Task information** form

- Select one or more tasks that you want to assign to a different calendar by using Shift-click or Ctrl-click.
- Open the **Task Information** form when selecting a single task by double-clicking on a task.
- Open the **Task Information** form when selecting multiple tasks by:
  - ➢ Select **Task**, **Properties** group, **Information**, or
  - ➢ **Shift+F2**, or
  - ➢ Right-click and select **Task Information**.
- Then select the **Advanced** tab:

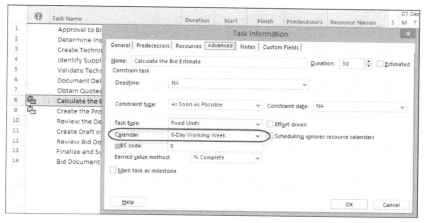

- From the **Calendar:** drop-down box select the calendar you want to assign to the task or tasks.

6.12.2 Assigning a Calendar Using a Column

You may also display the **Task Calendar** column and edit the task calendar from this column. The process of displaying a column is covered in the **FORMATTING THE DISPLAY** chapter. After a calendar has been assigned, an icon will appear in the **Indicators** column as displayed beside task 8 and 9 below:

The copy and paste cell or **Fill Down** command are useful for assigning Calendars from columns.

6.13   Workshop 4 - Adding Tasks

Background

If you do not have the default Microsoft Project settings installed on your computer, or you are not using the Eastwood Harris template, then you may not have the same results as displayed in these workshops.

It is simpler to teach Microsoft Project by showing how to enter the tasks first and then creating the summary tasks to represent the WBS Nodes or Products or Deliverables. Once a user understands the process, then the tasks and summary tasks and detailed tasks may be entered in any order.

Assignment

1. We will assume that the Planning Process is complete and we are to produce a schedule with several tasks for each Product (Deliverable) in the Work Breakdown Structure - WBS.

2. Use the columns to enter the Name and Duration of the tasks as below.

    ➤ The Estimate and Schedule will be completed by site personnel who work 6 days/ week.

    ➤ Assign the 6-Day Working Week calendar using the **Task Information** form **Advanced** tab to Tasks 8 and Task 9.

    ➤ Double-click on the task to open this form and select the **Advanced** tab, **Calendar** drop-down box to assign the Task Calendar.

3. A task will become a milestone when assigned a zero duration.

ID	Task Name	Duration	Task Calendar
1	Approval to Bid	0 days	
2	Determine Installation Requirements	4 days	
3	Create Technical Specification	5 days	
4	Identify Supplier Components	2 days	
5	Validate Technical Specification	2 days	
6	Document Delivery Methodology	4 days	
7	Obtain Quotes from Suppliers	8 days	
8	Calculate the Bid Estimate	3 days	Assign the 6 Days Working Week calendar
9	Create the Project Schedule	3 days	Assign the 6 Days Working Week calendar
10	Review the Delivery Plan	1 day	
11	Create Draft of Bid Document	6 days	
12	Review Bid Document	4 days	
13	Finalize and Submit Bid Document	2 days	
14	Bid Document Submitted	0 days	

Continued…

4. Your schedule should look like this using the Eastwood Harris Template:

	ⓘ	Task Name	Dur	Start	Finish	Total Slack	3 Dec '18	10 Dec '18
1		Approval to Bid	0d	3 Dec '18	3 Dec '18	8d		
2		Determine Installation Requirements	4d	3 Dec '18	6 Dec '18	4d		
3		Create Technical Specification	5d	3 Dec '18	7 Dec '18	3d		
4		Identify Supplier Components	2d	3 Dec '18	4 Dec '18	6d		
5		Validate Technical Specification	2d	3 Dec '18	4 Dec '18	6d		
6		Document Delivery Methodology	4d	3 Dec '18	6 Dec '18	4d		
7		Obtain Quotes from Suppliers	8d	3 Dec '18	12 Dec '18	0d		
8	📑	Calculate the Bid Estimate	3d	3 Dec '18	5 Dec '18	6d		
9	📑	Create the Project Schedule	3d	3 Dec '18	5 Dec '18	6d		
10		Review the Delivery Plan	1d	3 Dec '18	3 Dec '18	7d		
11		Create Draft of Bid Document	6d	3 Dec '18	10 Dec '18	2d		
12		Review Bid Document	4d	3 Dec '18	6 Dec '18	4d		
13		Finalize and Submit Bid Document	2d	3 Dec '18	4 Dec '18	6d		
14		Bid Document Submitted	0d	3 Dec '18	3 Dec '18	8d		

NOTES:

➢ The icon in the Information column on the left-hand side indicates that Tasks 8 and 9 have a non-standard calendar, which is the **6-Day Working Week** calendar set in this workshop.

➢ The Eastwood Harris template does not display the **Task Mode** column.

5. All tasks should be assigned the **Task Mode** of **Auto Scheduled**. If the task bars are not a solid blue color, except Task 7, which is critical and should be a solid red color, then they are **Manually Scheduled** and you have not set the options correctly in the Project Options.

6. Save your **OzBuild Bid** project.

7   ORGANIZING TASKS USING OUTLINING

7.1   Understanding Outlining

Outlining is used to summarize and group tasks under a hierarchy of **Parent** or **Summary Tasks**. They are used to add structure to your project during planning, scheduling and updating. These headings are normally based on your project WBS.

Defining the project's breakdown structure can be a major task for project managers. The establishment of templates makes this operation simpler because a standard breakdown is predefined and does not have to be typed in for each new project.

Projects should be broken into manageable areas by using a structure based on a breakdown of the project phases, stages, deliverables, systematic functions, disciplines or areas of work. The Outline structure created in your project should reflect the primary breakdown of your project, your WBS.

Microsoft Project 2000 introduced a new feature titled **Grouping**, which is similar to the **Group and Sort** function found in Oracle Primavera and Asta software. This feature allows the grouping of tasks under headings other than the "Outline Structure." **Grouping** is not the primary method of organizing tasks and is covered in the **TABLES AND GROUPING TASKS** chapter.

7.2   Creating an Outline

To create an **Outline**, a summary task needs to be created above the detail tasks. This may be achieved two ways:

- Using the **Insert Summary Task** (New to 2010) function by:
  - ➤ Selecting all the tasks, **NOT JUST ONE TASK**, you wish to create a summary task above,

  - ➤ Click on **Task**, **Insert** group, [⤵ Summary] button,
  - ➤ Rename the newly created Summary task:

- The more traditional way is by:
  - ➤ Inserting a new **Summary task** above the **detail tasks** and
  - ➤ Then **Demoting** the **Detail tasks** below **Summary task**. Demoting is explained in the next section.

The Start and Finish dates of the **Summary task** are adopted from the earliest start date and latest finish date of the **Detail tasks**.

The duration of the **Summary** is calculated from the adopted start and finish dates over the **Summary** task calendar, which is by default the **Project Calendar**.

7.3 Promoting and Demoting Tasks

Demoting or **Indenting** tasks may be achieved in a number of ways. Select the task or tasks you want to **Demote**. You may use any of the following methods to **Demote a selected task**:

- Click on the Indent ⊡ button on the **Task**, **Schedule** group, or

- Hold down the **Alt** and **Shift keys** and press the **Right Arrow Key** on your keyboard, or

- Move the mouse until you see a double-headed horizontal arrow in the task name field, left-click and drag the task right. A vertical line (see lower of the two pictures below) will appear indicating the outline level you have dragged the task(s) to:

Before dragging tasks to indent:

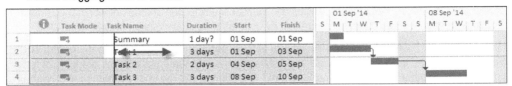

After dragging tasks to indent:

Promoting or **Outdenting tasks** uses the same principle as demoting tasks. Select the task or tasks you want to promote, ensure you have selected the whole task and not just some cells, then you may:

- Click on the Outdent ⊡ button, or

- Add the ⊡ button to the **Quick Access Toolbar...** or download the **Eastwood Harris Quick Access Toolbar**, or

- Hold down the **Alt** and **Shift keys** together and press the **Left Arrow Key** on your keyboard, or

- Move the mouse until you see a double-headed horizontal arrow in the task name column, left-click and drag the tasks left.

Tasks may be added under a **Detail task** and demoted to a third level and so on.

 It is considered good practice to remove the link from "Task 2" to "Task 3" and replace it with a link from "Task 2" to "Task 3.1" thus not having any relationships between summary tasks.

7.4   Summary Task Duration Calculation

The summary task duration is calculated from the Start to the Finish over the calendar assigned to the task, thus changing the summary task calendar will change the displayed duration:

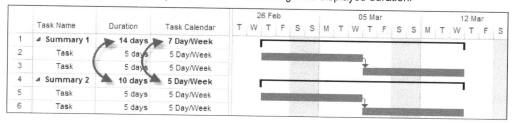

7.5   Summarizing Tasks

Once you have created summary tasks, the detail tasks may be rolled up or summarized under the summary tasks. Rolled up tasks are symbolized by the ▷ sign to the left of the summarized task description:

- This picture shows **Detail Task 3** rolled up.

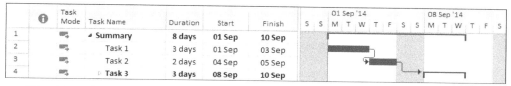

- This picture shows a **Summary Task** rolled up.

7.5.1   Summarize and Expand Summary Tasks to Show and Hide Subtasks

Select the task you want to hide:

- Click on the ◢ to the left of the Task Name, or

- Click on the **Hide Subtasks** ▭ button that may be added to the Quick Access Toolbar, or

- Double-click on the **Task ID**.

Displaying rolled-up tasks is similar to rolling them up. To do this, select the task you want to expand. Then:

- Click on the ▷ to the left of the Task Name, or

- Click on the **Show Subtasks** button that may be added to the Quick Access Toolbar, or

- Double-click on the **Task ID** (not the **Task Name** as this will open the **Task** form).

7.5.2 Roll Up All Tasks to an Outline Level

A schedule may be rolled up to any Outline Level by selecting the desired Outline Level from the **View**, **Data** group, Outline button:

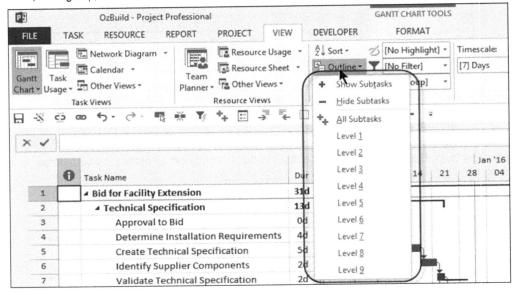

7.6 Project Summary Task

A **Project Summary Task** may be displayed by checking the **Format**, **Show/Hide** group, **Project Summary Task**. This task spans from the first to the last task in the project and is in effect a built-in Level 1 outline task. The description of the summary task is the Project Title entered in the **File**, **Info**, **Project Information**, **Advanced Properties** form. A Project summary task is a virtual task and may not have resources, relationships or constraints assigned, and is assigned a Task ID of 0.

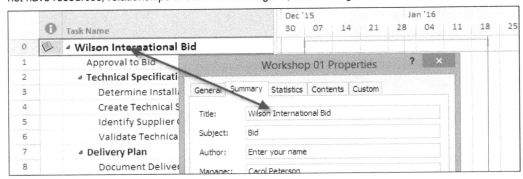

7.7   *Workshop 5 - Entering Summary Tasks*

Background

The summary tasks may be used to represent a Project, Deliverables/Products, WBS Nodes, Phase, Stages or Work Packages.

We will add summary tasks to represent the Initiation Phase and Deliverables/Products of the Initiation Phase.

Assignment

1. Display the **Project Summary Task** by checking the **Format**, **Show/Hide** group, **Project Summary Task** and close the form.

2. Observe how the **Project Summary Task** is formatted, it will be **Bold** and have a Task ID of "0".

3. Hide the **Project Summary Task** by repeating para 1 and unchecking, as we will create an Outline Level for the Project summary task.

4. Create an Outline Level 1 for Phase entitled **Bid For Facility Extension** and

5. Create an Outline Level 2 for each of the three Products:

    - **Technical Specification**

    - **Delivery Plan**

    - **Bid Document**

6. Try using the various methods for indenting and outdenting tasks.

7. Your schedule should look like this with the Eastwood Harris template:

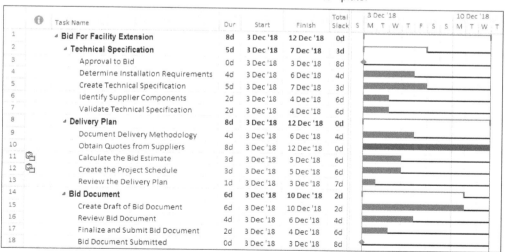

8. Save your **OzBuild Bid** project.

---

8 FORMATTING THE DISPLAY

This chapter covers the following topics, which are used to format the on-screen display which are also reflected in print preview and printouts:

Topic	Menu Command
Inserting Columns	• Highlight a column and strike the **Ins Key**, or • Select **Format**, **Columns** group, **Insert Column**, or • Right-click and select **Insert Column**.
Deleting Columns	• Highlight a column and strike the **Delete** key, or • Select **Format**, **Columns** group, **Column Settings**, **Hide Column**, or • Right-click and select **Hide Column**.
Adjusting the Width and Moving Columns	• Grab the right header border line with the mouse and drag or resize. • Double click on the right hand edge of the column header to optimize the header size.
Table – formatting the columns of data	• **View**, **Data** group, **Tables**, 📋 **More tables…** or • Select a column and right-click to insert, or • Right-click on a column header and select **Field Settings**.
Formatting One Column	• Right-click on the column title and select **Field Settings**.
Format Bars	• Open the Bar Styles form by left clicking in the Gantt Chart, or • **Format**, **Bar Styles** group, or • **Format**, **Gantt Chart Styles** group, or • The **Gantt Chart Wizard**. This should only be used with projects created with Microsoft Project 2010 and earlier.
Formatting Time Units	• Select **File**, **Options**, **Advanced**, **Default options for this project:**
Row Height	• Drag one or more selected rows with the mouse, or • Edit the Table, or • **Wrap Text** command found on the **Format**, **Columns** group, **Column Settings**, 📋 **Wrap Text** which automatically adjusts the row height to fit the text into the available column width. Ensure you select the column when turning on or off this function.
Timescale	• **Zoom Slider** at the bottom right-hand side of the screen, or • Double-click on the timescale opening the **Timescale** form, or • Use the 🔍 🔍 **Zoom** buttons.
Gridlines	• **Format**, **Format** group, **Gridlines**, **Gridlines…**
Relationship Lines	• **Format**, **Format** group, **Layout**
Format Text Font	• Select the text to be formatted and right-click.

The formatting is applied to the current **View** only and is automatically saved as part of the View when another View is selected. Views are covered in more detail in the **VIEWS AND DETAILS** chapter.

8.1 Formatting the Columns

Microsoft Project has some column formatting functions which are intended to make it simple to add and format new columns. There are two methods of formatting the columns:

- Inserting, editing and deleting columns of data using the **Column Definition** form.

- **Tables** – A table may be created or an existing table edited with required columns using the **Table Definition** form. You may set up the data columns in the way you want to see the information on the screen and in printouts. Therefore **Tables** may be created, edited and deleted and you may select which one is used to display the data with each View. A Table may be assigned to multiple Views.

 As time progresses with the option **File**, **Options**, **Advanced**, **Automatically add new views**, **tables**, **filters**, **and groups to the global** activated, a project that is created from the **Blank Project** will have many Views, Tables and Filters from old projects that may be irrelevant to the current project and it is suggested this be turned off.

8.1.1 Understanding Custom Fields

A Custom Field is a task or resource field that may be renamed and user defined data entered in columns. These fields are preformatted only to accept specific data such as dates, costs, durations or text. These are covered in more detail in the **TABLE AND GROUPING TASKS** chapter.

8.1.2 Column Names

Some of the Microsoft Project column names are confusing or difficult to find and the table below identifies some of the more common names and what they are to enable you to find columns more quickly:

Microsoft Project Field Name	Common Name	Location
Name	Task Name	Gantt Chart
Task Calendar	Calendar	Gantt Chart
Task Start	Start	Bars form
Task Finish	Finish	Bars form

8.1.3 Inserting Columns

Insert a column by clicking on the column title where you require the new column. This will highlight the column. To insert a new column:

- Select **Format**, **Columns** group, **Insert**, **Column...** ⬚, or
- Hit the **Ins** Key, or
- Right-click and select **Insert Column...** ⬚.

Then select the column from the drop-down list. –

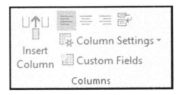

- You may immediately start typing, which will take you to the appropriate position in the list.

- In Microsoft Project 2013 and 2016, when a column name that does not exist is typed into the header, then an existing Text column will be renamed with the new title. Therefore the inserted column is a renamed text **Custom Field**.

8.1.4 Format Columns Group

The **Format**, **Columns** group has the following functions:

- 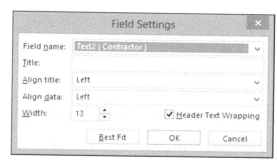 – Aligns the text to the left, center or right,

- **Wrap Text** and increases the row height so all text is visible,

- **Column Settings** has:

  > ✕ – **Hide Column** which hides a column but does not delete the data.

  > – **Field Settings** that opens the **Field Settings** form:
  > **NOTE:** The picture displays **Text2 Custom Field** that has been renamed **Contractor**.

  > – **Data Type** which allows the data type of a Custom Field to be changed.

  > – **Display Add New Column** is a function that was new to Microsoft Project 2010. A column may be permanently displayed on the right-hand side of the screen titled **Add New Column** and clicking on this column will open up a drop-down box for the selection of the data type. This option will display or hide this column.

  > – **Insert Column** inserts a new column.

- **Custom Fields** opens the **Custom Fields** form covered in the **TABLE AND GROUPING TASKS** chapter.

8.1.5 Hiding Columns

Hiding a column does not delete the data as in Excel; the software is just not reading the data from the database. Hiding a column may be achieved by highlighting the column or by clicking on the title and then:

- Select **Column Settings**, ✕ **Hide Column**,

- Hit the **Delete** key, or

- Right-click and select ✕ **Hide Column**.

1. There will be no confirmation of hiding a column, but you are allowed to undo the hiding.

2. The term "hide" means the column is removed from the Table in this project but the data is not deleted as in Excel when a column is deleted.

8.1.6 Adjusting the Width of Columns

You may adjust the width of the column either manually or automatically.

- For manual adjustment, move the mouse pointer to the right vertical line of the column in the header. A ⟷ mouse arrow will then appear and enable the column to be adjusted.

Task Name	Duration	Start	Finish
Task 1	3days	01 Jan	03 Jan
Task 2	2days	06 Jan	07 Jan

- For automatic adjustment, once again position the mouse pointer to the right vertical line of the column in the header and double left-click the mouse. The column width will automatically adjust to the best fit.

8.1.7 Moving Columns

Columns in a Table may be moved by clicking on the column header. The mouse pointer will change

to a 🔼 and the column may be dragged to a new location.

8.1.8 Formatting Columns Using the Table Function

- Select **View**, **Data** group, **Tables**, and select from the list of predefined **Tables** listed on the menu. Select the table you want to display:

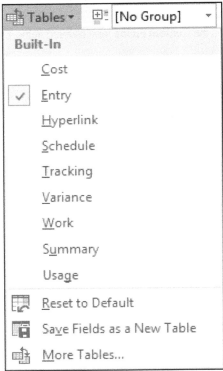

- Select **View**, **Data** group, **Tables**, **More Tables…** to open the **More Tables** form:

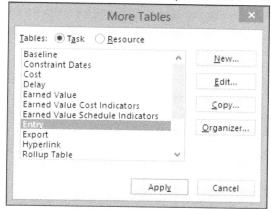

- ➢ <u>New…</u> – To create a new Table.

- ➢ <u>Edit…</u> – To edit the highlighted Table.

- ➢ <u>Copy…</u> – To copy the highlighted Table.

- ➢ <u>Organizer…</u> – Opens the **Organizer** form which enables you to copy a Table from one opened project to another or to the Global Project.

- ➢ <u>Apply</u> – Applies the selected Table making it visible on the screen.

- When you select <u>New…</u>, <u>Edit…</u> or <u>Copy…</u> you will be presented with the **Table Definition** form:

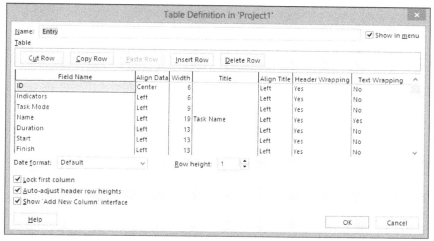

- ➢ Click on the **Show in menu** box to display the Table in the **View**, **Data** group, **Tables** menu.

- ➢ The columns of data will be displayed on screen from left to right in the same order as the rows in the form.

- ➢ Highlight a row and then you may use the <u>Cut Row</u>, <u>Copy Row</u>, <u>Paste Row</u>, <u>Insert Row</u> or <u>Delete Row</u> buttons.

- ➢ The data to be displayed may be selected from the drop-down box in the **Field Name** column.

- ➢ **Align Data** and **Width** are used for formatting the data in the columns.

> ➢ The Microsoft Project **Field Name** may be replaced by typing your own title in the **Title** box.

> ➢ The **Date format:** drop-down box is used to change the format for this table only.

 This is a very useful function to ensure that other users of the project file see the intended date format and not their system default date format.

- **Row Height:** sets the default height of all the rows in this table. A row height may be changed by dragging the cell boundary line once a task has been created.

  > ➢ **Lock first column** prevents the first column from scrolling and is useful when the first column contains the Task Name.

  > ➢ **Auto-adjust header row heights** will automatically adjust header row heights to the width of the column.

  > ➢ **Show 'Add New Column' interface** hides or displays the **Add New Column** column (new function to Microsoft Project 20103) on the right-hand side of the columns in all Views, used for adding new columns.

To save a table for use in all your new projects, copy the table to the **Global.mpt** template using **File**, **Info**, **Organizer** and select the **Tables** tab. This will not copy the new names of renamed fields.

You may also copy a **Table** to another open project or rename a **Table** using **File**, **Info**, **Organizer** and selecting the **Tables** tab.

 Applying a Table to a View will permanently change the View unless the file is not saved. The Gantt Chart View has the Entry Table assigned by default.

8.2 Formatting Time Units

Select **File**, **Options**, **Advanced**, **Display options for this project:**

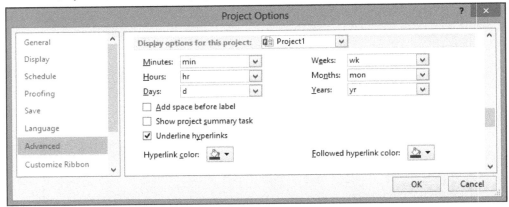

- The **Display options for this project:** always specifies the time units, for example **day**, **dy** or **d**.

- Uncheck the **Add space before label** check box to remove a space between the value and label in date columns which allows a narrower Duration column to be displayed.

 To make the Duration column width narrow, **Days:** should be set to **d** and the **Add space before label** unchecked. The column header could also be edited to **Dur** to ensure the header is also narrow. This will provide more space for other data:

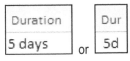

8.3 Formatting the Bars

Microsoft Project has several options for bar formatting:

- All the bars may be formatted to suit user definable parameters, or

- Individual bars may be formatted.

This section will cover the formatting of all the bars and the next section will cover formatting individual bars.

Most formatting only affects the current View.

All bars in the Gantt Chart may be formatted to suit your requirements for display by:

- Opening the **Bar Styles** form by double-clicking anywhere in the Gantt Chart area, or

- **Format**, **Bar Styles** group, or

- **Format**, **Gantt Chart Styles** group, or

- The **Gantt Chart Wizard**, which is a simple way to format bars.

 As in Microsoft Project 2010, many of the formatting menu options in Microsoft Project 2013 and 2016 are designed for a schedule created from a Microsoft Project 2013 or 2016 template. Some formatting menu options will not operate as expected if you have opened a project created in an earlier version of Microsoft Project. For example, when a Microsoft Project 2007 file is opened it with Microsoft Project 2013, the author found the display Baseline bar found on the Ribbon command **Format**, **Bar styles**, **Baseline** was incompatible with projects created in earlier versions of Microsoft Project and resulted in the Baseline bar being drawn over the Current Schedule bar. Users may have to use the manual method of formatting bars by opening the bars form or use the **Gantt Chart Wizard** to format bars created in earlier versions of Microsoft Project.

8.3.1 Formatting All Task Bars Using the Bar Styles Form

To format all the bars you must open the **Bar Style** form by:

- Double-click anywhere in the Gantt Chart area, but not on an existing bar, as this will open the **Format Bar** form for formatting an individual bar, or

- Select **Format**, **Bar Styles** group, **Format, Bar Styles**, the picture below displays the default **Bar Styles** when Microsoft Project is loaded:

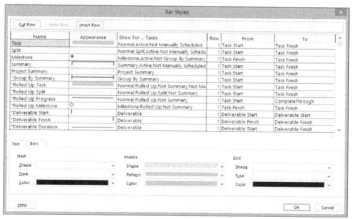

 The picture above shows a typical default Microsoft Project Bar Styles setting from a default load of Microsoft project.

The following notes are the main points for using this function. Detailed information is available in the Help facility by searching for "Bar styles dialog box."

- Each bar listed in the table will be displayed on the bar chart.

- Bars may be deleted with the ⎡ Cut Row ⎤ button, pasted using the ⎡ Paste Row ⎤ button and new bars inserted using the ⎡ Insert Row ⎤ button.

- The **Name** is the title you may assign to the bar and is displayed in the printout legend. To hide the bar on the legend precede the **Name** with an *. There are many bars with an * by them as default, as displayed in the picture above.

- The appearance of each bar is edited in the lower half of the form. The bar's start point, middle and end points may have their color, shape, pattern, etc. formatted.

- When a new Milestone is created, the **From** and **To** must both be set to **Task Finish**.

- **Show For ... Tasks** allows you to select which tasks are displayed, similar to a filter. More than one task type may be displayed by separating each type with a ",". Bar types not required are prefixed with "**Not**." For example, the ⎡Normal,Rolled Up,Split,Not Summary⎤ bar would not display a bar for a summary task. Should you leave this cell blank then all task types will be displayed in this format.

 The bars may be placed on one of four rows numbered from 1 to 4, top to bottom. If multiple bars are placed on the same row then the bar at the top of the list will be drawn first and the ones lower down the list will be drawn over the top, thus potentially hiding the ones below in the list.

- **From** and **To** allow you to establish where the bars start and finish. The picture below shows how to format **Total Float**, **Free Float** and **Negative Float**. Unlike some other planning and scheduling software, the Negative Float is drawn from the Start Date of a task and not the Finish Date and therefore a separate bar is required for Negative and Positive Float.

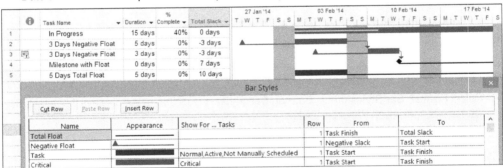

 By default Microsoft project only displays the **Free Float Bar** with the **Format**, **Bar Styles** group, **Slack** button not the **Total Float** bar that would normally be expected.

Also the Negative Float bar is not automatically displayed by any Microsoft Project function, nor is it included in any View. These are two of the most important bars to show when a project finish date has been set using a constraint and these bars must always be manually created. There are some options to resolve this:

- You may wish to consider recording a macro to create these bars and this can be run when a negative float bar is required. Recording a macro is covered in para 25.15.

- The author has found that these bars produce a better presentation and do not interfere with the drawing of relationships when created at the bottom of the list in the **Bar Styles** form.

- You may download an Eastwood Harris Microsoft Project 2013 and 2016 template project from the Eastwood Harris web site at www.eh.com.au **SOFTWARE AND DOWNLOADS** page. This has an inbuilt View which displays both the Total and Negative Float bars. Also other issues with Microsoft Project 2013 and 2016 have been resolved and are covered in the template description.

- The **Text** tab allows you to place text inside or around the bar:

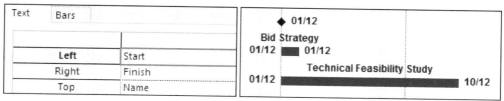

- It is not possible to format the font in the **Bar Styles** form. Select **Format**, **Format** group **Text Styles** to open the **Text Styles** form and select the bar text font.

To show Critical and Non-critical tasks, the bars should be formatted as shown below, with particular attention paid to the **Show For ... Tasks** column. Non-critical Tasks are formatted as **Normal**, **Non-critical** and Critical Tasks as **Normal**, **Critical**.

Name	Appearance	Show For ... Tasks	Row	From	To
Task		Normal,Noncritical,Active,Not Manually S	1	Task Start	Task Finish
Critical		Normal,Critical,Active,Not Placeholder	1	Task Start	Task Finish

8.3.2 Format Bar Styles Group Menu

Select the **Format**, **Bar Styles** group to view this group of commands:

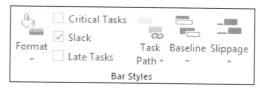

- **Format** has two options:

  ➢ **Bar** formats of one or more selected bars and is covered in the next section.

  ➢ **Bar Styles** opens the **Bar Styles** form as discussed above to format all bars.

- The other buttons in the **Format**, **Bar Styles** groups will hide or display the bars as indicated, but these may not give the expected results with a project created in an earlier version of Microsoft Project.

  ➢ **Critical Tasks** shades the Critical Tasks red.

  ➢ **Slack** displays the Free Float and **NOT** Total Float bar which were displayed in earlier versions on Microsoft Project when the Gant Chart wizard was run.

  ➢ **Late Tasks** displays tasks that are late compared to the **Status Date**.

  ➢ **Baseline** displays the Baseline bar.

  ➢ **Slippage** displays how much time the task is behind the Baseline.

⚠ Many of the formatting menu options are designed for a schedule created from a Microsoft Project 2013 template. Some menu options will not operate as expected if you have opened a project created in an earlier version of Microsoft Project. For example, the author found the baseline bar was placed on top of the current schedule bar using the Ribbon commands. Users may have to use the manual method of formatting bars by opening the bars form or using the **Gantt Chart Wizard**.

8.3.3 Gantt Chart Styles Group Menu

Select **Format**, **Gantt Chart Styles** group to see the option for coloring bars. This function was new to Microsoft Project 2010.

The button at the bottom right-hand side opens the **Bar Styles** form.

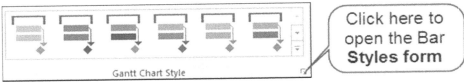

 See warning above.

8.3.4 Formatting Bars Using the Gantt Chart Wizard

The Gantt Chart Wizard is a popular function for people who used earlier versions of Microsoft Project. The wizard will overwrite any formatting you may have created. This is a straightforward method of formatting your bars and often this is the best method of formatting bars. It is very simple to use but will not display the **Negative Float** and **Free Float** bars. These will have to be added manually using the **Bar Style** form as described in the **Bar Styles** form section.

 If the **Gantt Chart Wizard** button is not on your Microsoft Project default toolbars then this may be added to either the **Quick Access Toolbar** or the **Ribbon**.

The author found that projects formatted with the **Gantt Chart Wizard** and projects created in earlier versions of Microsoft Project will have formatting that is incompatible with the Microsoft Project 2013 and 2016 **Format**, **Bar Styles** group buttons. The use of the **Baseline** button resulted in the Baseline bar covering the current schedule bar so only the Baseline bar was visible. To resolve this issue, users should either use only **Gantt Chart Wizard** or only the new Microsoft Project 2013 and 2016 **Format**, **Bar Styles** group buttons, but not both. This issue may be rectified with software updates.

8.3.5 Placing Dates and Names on Bars

To place a name and or dates on the **Task bars** and **Milestones** create two bars at the bottom if the list in the bars form as per the picture below which neither displays a bar:

Name	Appearance	Show For ... Tasks	Row	From	To
Task Name on Bar		Not Milestone	1	Task Start	Task Finish
Task Name on Milestone		Milestone	1	Task Finish	Task Finish

Text   Bars

Left	Start	
Right	Finish	
Top	Name	

No Bar is displayed

From and To must be **Task Finish** for the Milestone

8.4   Row Height

8.4.1   Setting Row Heights

Row heights may be adjusted to display text that would otherwise be truncated by a narrow column.

Row heights are adjusted by whole lines and not points as in Excel.

The row height may be set in the **Table Definition** form by selecting **View**, **Data** group **Tables**, **More Tables....** From this view select the table in which you wish to edit the row height in and click on the <u>Edit...</u> button. Once the **Table Definition** form is open select the row height from the drop-down box next to **Row height:**.

The row height of one or more columns may also be adjusted in a similar way to adjusting row heights in Excel, by clicking on the row and dragging with the mouse:

- Highlight one or more rows that need adjusting by dragging or Ctrl-clicking.

- If all the rows are to be adjusted, then click on the **Select All** button above row number 1, to highlight all the tasks.

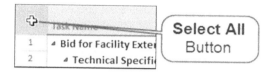

- Then move the mouse pointer to the left-hand side of a horizontal row divider line. The pointer will change to a double-headed arrow ↨. Click and hold with the left mouse button and drag the row or rows to the required height.

8.4.2   Wrap Text Command

Microsoft Project 2010 has introduced a **Wrap Text** command found on the **Format**, **Columns** group, **Column Settings**, ⊞ **Wrap Text** which automatically adjusts the row height to fit the text into the available column width.

Ensure you select the column when turning on or off this function and when it is not highlighted it is turned off.

8.5   Format Fonts

There are two basic options for formatting fonts:

- Either individual cells may be selected by Ctrl-clicking or dragging with the mouse and formatted, or

- The fonts of tasks that meet pre-set criteria, such critical tasks may be formatted using the text **Styles** command.

8.5.1   Format Individual Cells Font Command

The **Format**, **Font…** function from Microsoft 2007 and earlier has been replaced with a right-click option and allows you to format any selected text in selected cells, rows or columns:

- Select the text to be formatted,

- Right-click and two toolbars are opened. The upper toolbar in the picture below has four buttons that may be used for formatting individual cells:

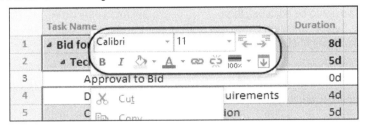

8.5.2   Format Text Styles

The **Format**, **Format** group, **Text Styles** command opens the **Text Styles** form and allows you to select a text type from the **Item to Change:** drop-down box and apply formatting to the selected item:

Text may be formatted by using any of the styles listed below:

- **All**: This is all text including columns and rows,

- **Non-critical**, **Critical**, **Milestone**, **Summary**, **Project Summary**, **Marked**, **Highlighted** and **External tasks**,

- **Row** and **Column** titles,

- **Top**, **Middle** and **Bottom Timescale Tiers**, and

- **Bar Text** left, right, below, above and inside.

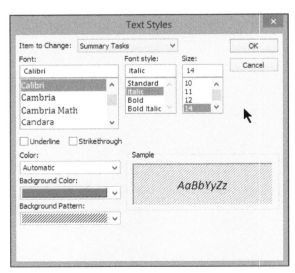

8.6   Format Timescale

8.6.1   Zoom Slider

The **Zoom Slider** was introduced with Microsoft Project 2010 and replaced the <u>V</u>iew, <u>Z</u>oom... function. This may be found at the bottom right-hand side of the screen and provides a simple way of scaling the time scale in the Gantt Chart and all other time scaled views such as the Calendar, Usage and Network Diagram View.

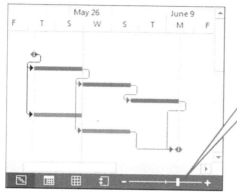

Dragging the **Zoom Slider** to zoom the timescale is not recommended, read the comments below

 This function works differently than other scheduling software in that it changes the scale and the displayed time units at the same time and may result in some undesirable time units being displayed.

Once this function is used, your original timescale date formatting will be lost and may only be recovered with undo as this function applies its own formatting such as date formats.

The author has found that more predictable results are achieved by using the traditional ⊕ ⊖ **Zoom In** and **Zoom Out** functions which may be added to the **Quick Access Toolbar**.

8.6.2   Ribbon Menu

There are some new commands available with Microsoft Project 2010 on the **Ribbon**:

- The **Timescale:** option is a quick method of selecting the **Minor Timescale**, the lower line in the Timescale,

- Zoom opens a self-explanatory menu,

- **Entire Project** zooms the timescale to fit the whole project Gantt Chart to fit in the available space,

- **Selected Tasks** zooms the timescale to fit the bars of selected tasks to fit in the available space,

---

8.6.3   Format Timescale Command

The **Timescale** form provides a number of options for timescale display, which is located above the Bar Chart, and the shading of **Nonworking** time.

To open the **Timescale** form:

- Double-click on the timescale, or
- Add the **Timescale** button to the Ribbon or Quick Access Toolbar.

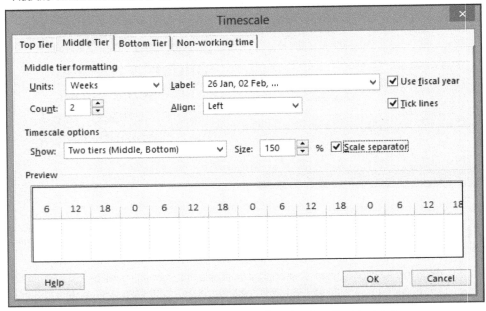

There are many options here which are intuitive and will not be described in detail.

Top Tier, Middle Tier and Bottom Tier Tab

- These three timescales may have different scales. These are often set at "weeks and days" or "months and weeks." By default, the Top Tier timescale has been disabled. You may enable the three tiers together by selecting Three Tiers (Top, Middle, Bottom) from the **Timescale options, Show:**.

- The **Label** will affect how much space the timescale will occupy, so the selection of a long label will result in longer Task bars.

- **Tick lines** and **Scale separator** hide and display the lines between the text.

- **Size:** controls the horizontal scale of the timescale and in association with the **Label:** are the two main tools for scaling the horizontal axis in the Gantt Chart.

- Choose the **Use fiscal year** function to display the financial year and then select the **File, Options, Calendar** tab to choose the month in which the fiscal year starts.

- Should you wish to number the time periods, for example; **Week 1**, **Week 2**, etc., there are a number of sequential numbering options available at the bottom of the label list.

Nonworking Time Tab

The **Nonworking time** tab allows you to format how the nonworking time is displayed. You may select only one calendar. The nonworking time may be presented as shading behind the bars, in front of the bars or hidden.

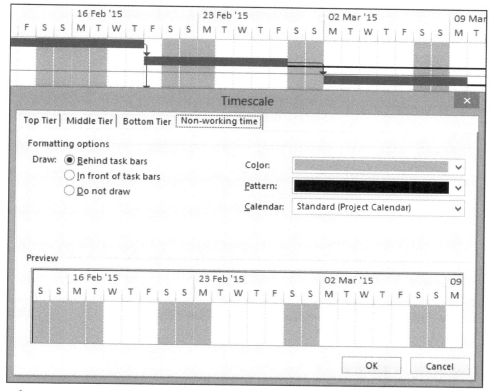

 By default this is set to the Standard Calendar for each view and does not change when the Project Default Calendar is changed.

Therefore if you change the Project Base calendar in the **Project Information** and you wish to see this new calendar in all views you will have to edit all the views.

8.6.4   Format Timescale Font

To format the Timescale font, the **Format**, **Format** group, **Text Styles** command opens the **Text Styles** form:

The timescale fonts may be formatted separately by selecting the appropriate line item under **Item to Change:**

A very tight timescale may be achieved by making the Bottom Timescale Tier a small font as displayed in the picture.

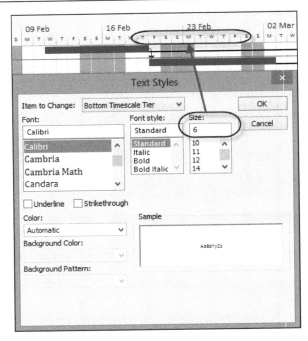

8.7   Format Gridlines

Gridlines are important to help divide the visual presentation of the Bar Chart. This example shows **Middle Tier Gridlines** every week and **Bottom Tier Gridlines** every day.

To format the Gridline select **Format**, **Format** group, **Gridlines**, ⊞ **Gridlines…** or **Right click** in the **Gant Chart** and select **Gridlines** to open the **Gridlines** form:

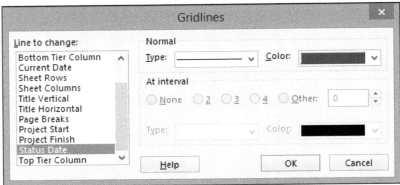

- Select the gridline from the drop-down box under **Line to change:**
- Select color and type from under **Normal**.
- Date gridlines may be set to occur at intervals using the **At interval** option.

Some of the titles for the gridlines are not intuitive, so some interpretation is given below:

- For Data Column and Row dividing lines, use **Sheet Rows** and **Sheet Columns**.
- For Timescale and Column Titles, use **Horizontal** and **Title Vertical**.
- Gantt Chart area, including lines for **Project Start** and **Finish Date**, **Current** and **Status Date**, are clearly described.
- **Page Breaks** will only display manually-inserted breaks. You may need to add the **Insert Page Break** [icon] **Button** on the **Quick Access** Toolbar.

 The earlier Microsoft Project option **Manual page breaks** check box, which allowed printing and ignoring manual page breaks, in the **Print** form has been removed from Microsoft Project 2013.

Microsoft Project has two dates that may be used to identify the **Data Date**, the date that the data has been collected for updating a project schedule. These two dates are the **Status Date** and **Current Date** which are set in the **Project Information** form:

- By default Microsoft Project displays the **Current Date** as a dark dotted vertical line but this is reset to the computer's system date each time the project file is opened. It is suggested that this line be removed using the **Gridline** form and the **Current Date** is not used to identify the **Data Date** because the software changes it every time the schedule is opened.
- The **Status Date** never changes once set and therefore it is suggested that this line should be displayed as per the picture above.

 Many laser printers will not print light gray lines clearly, so it is often better to use dark gray or black Sight Lines for better output.

8.8   Format Colors

Colors are formatted in a number of forms and there is no single form for formatting all colors:

- **Nonworking time** colors in the Gantt Chart are formatted in the **Timescale** form, double-click on the timescale.
- **Text** colors are formatted in the **Text Styles** and **Font** forms, found in the **Format**, **Format** groups, **Text Style**.
- **Gridline** colors are formatted in the **Gridlines** form, also found under the **Format**, **Format** group, **Gridlines**.
- **Hyperlink** colors are formatted under **File**, **Options**, **Advanced**, **Display options for this project:**
- **Timescale** colors are with the **File**, **Options**, **General** tab, **User Interface Options**, **Color scheme:** option.
- The **Logic Lines**, also known as **Dependencies**, **Relationships**, or **Links**, inherit their color from the predecessor's bar color in the Gantt Chart view and may be formatted in the Network Diagram view by selecting **Format**, **Format** group, **Layout**.

8.9 Format Links, Dependencies, Relationships, or Logic Lines

The Links, also known as Dependencies, Relationships, or Logic Lines, may be displayed or hidden by using the **Layout** form.

- Select **Format**, **Format** group, **Layout** to open the **Layout** form and click on one of the three radio buttons under **Links** to select the style you require:

The color of the **Link** is inherited from the color of the predecessor task.

- To display Critical Path on the relationship lines you will need to format the bars with a different color. This is often set to red.

 The color of the successors' relationship lines is adopted from the task bar color. Therefore, re-formatting critical bars with the **Format Bar** form will also re-format the color of the successors' relationship lines and they may no longer display the Critical Path color on the Logic Lines. This will effectively mask the Critical Path and could provide misleading results.

The highest bar in the Bars form dictates the color of the relationship line and may not be the same as the color displayed in the Gantt Chart View.

Oracle Primavera products format the relationship separately from the bars and are able to identify the Critical, Driving and Non-Driving relationships, which is not possible with Microsoft Project.

8.10   Workshop 6 - Formatting the Bar Chart

Background

Management has received your draft report and requests some changes to the presentation.

If you are using the Eastwood Harris template then most of the formatting requirements you need to make are made in the template in the **Gant Chart Inc Total Float and Neg Float** view. The following attributes have been changed from the standard settings:

- **File**, **Options**, **Schedule** and **Advanced** have been edited in line with the Author's recommendations in his book.

- Two new Views titled **Gantt Chart Inc Negative and Total Float** and **Gantt Chart Name on Bars**

- **Columns:** Total Float added and Resources removed from the **Gant Chart Inc Total Float and Neg Float** view.

- **Grid lines:** Middle and Bottom Timescale Tiers, Project Start, Project Finish and Status Date displayed and Current Date removed.

- **Bars:**
  - ➢ Total Float (Total Slack) and Negative Float added and all text removed from all bars.
  - ➢ Bar display in the Legend: Many bars have been hidden in the Legend (but not deleted) by placing an "*" at the front of the Bar description in the Bars form.

- **Printing:** some project information is drawn from the Advanced Properties form. Also all project data has been removed from the Legend so the Legend may be hidden if not required, thus leaving all project data displayed if the Legend is hidden.

- A Custom Field has been added to the **Tracking Table** titled **Status Check** that indicates when activities have been updated correctly. NOTE: You must set the "Status Date" in the "Project Information" form for this field to calculate correctly.

Assignment

Format your schedule as follows:

1. Select **Task**, **Views**, **Gantt Chart**, **Custom** and select **Gant Chart Inc Total Float and Neg Float** to apply this view.

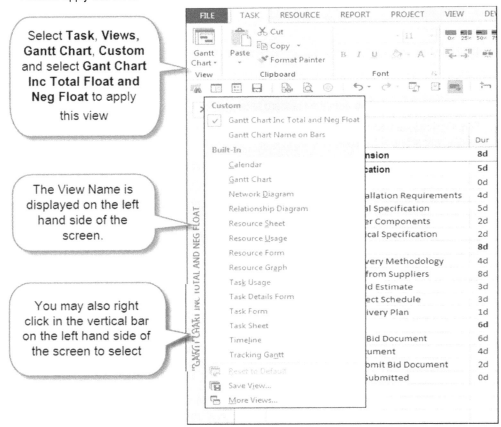

2. Then apply the **Gantt Chart Name on Bars** view and then the **Gantt Chart,**

3. Your answer show be as per the pictures below:

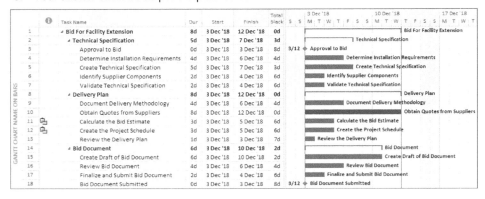

4. Apply the **Gant Chart** view.

5. Ensure that the **Entry Table** is displayed by selecting **View**, **Data** group, **Tables**, **Entry** table.

6. Apply the **Costs** table,

7. Then apply the **Tracking** table,

8. Reapplying the **Gant Chart Inc Total Float and Neg Float** view.

9. Insert the **Task Calendar** column between **Duration** and **Start** columns in the **Table Definition** form by using the right click command:

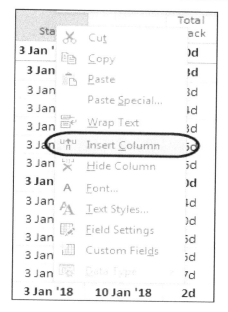

10. Your answer show be as per the pictures below:

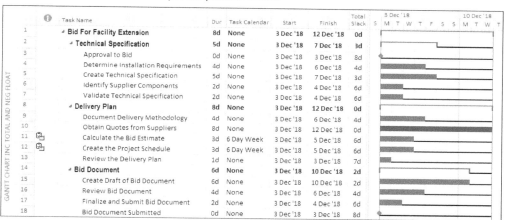

11. Add the 🔍 🔍 **Zoom In** and **Zoom Out** buttons to the **Quick Access Toolbar** if these icons are not present on the toolbar and test their function.

12. Leave the scaling at months and weeks with **Size** in the **Timescale** form of 150%.

13. Save your **OzBuild Bid** project.

---

9   ADDING TASK DEPENDENCIES

9.1   *Understanding Dependencies*

This chapter will concentrate on the basis that all tasks are **Auto Scheduled**. There is a section that covers the relationships between **Manually Scheduled** tasks covered in chapter 22.

The next phase of a schedule is to add logic to the tasks. There are two types of logic that you may use:

- **Dependencies** (**Relationships** or **Logic** or **Links**) between tasks, and
- Imposed **Constraints** to task start or finish dates. These are covered in the **CONSTRAINTS** chapter.

Microsoft Project's Help file and other text uses the terms "**Dependencies, Relationships and Links**" for Dependencies but does not use the term "**Logic**."

There are a number of methods of adding, editing and deleting task **Dependencies**. We will look at the following techniques in this chapter:

Topic	Notes for creating a Finish-to-Start Dependency
• Graphically in the Gantt, Calendar or Network Diagram Views	Drag the 4-headed mouse pointer ✛ from one task to another to create an FS dependency.
• With the **Task**, **Schedule** group **Link** and **Unlink** buttons	Select the tasks in the order they are to be linked and click on the Link ⊖ or ⊖ buttons.
• By using the Menu command	Select the tasks in the order they are to be linked and select **Ctrl+F2**.
• By opening the **Task Information** form	Predecessor only may be added and deleted.
• Through the **Predecessor** and **Successor Details** forms	Open the bottom pane, **Windows**, **Split** and then right-click and select **P**redecessors and **Successors**.
• By editing or deleting a dependency using the **Task Dependency** form	Double-click on a task link (relationship line) in the **Bar Chart** or **Network Diagram** view.
• **Autolink inserted or moved tasks**	Select **File**, **Options**, **Schedule** tab, **Scheduling options for this project:** section, and check the **A**utolink inserted or moved tasks box. This option is NOT recommended.
• By displaying the **Predecessor** or **Successor** column	Edit the relationships in the columns.

Microsoft Project allows only one relationship between two tasks whereas Oracle Primavera P6 allows four and Asta Powerproject allows an unlimited number of relationships between two activities.

Dependencies

Generally, there are two types of dependencies that may be entered into the software:

- **Hard Logic**, also referred to as **Mandatory Logic**, are dependencies that may not be avoided. For example a footing would have to be excavated before it may be filled with concrete or a computer and software delivered before the software may be installed onto the computer.

- **Soft Logic**, also referred to as **Sequencing Logic** or **Preferred Logic**, which often may be changed at a later date to reflect planning changes. An example would be determining the order in which a number of footings are dug or which computer is installed with software first.

To create a **Closed Network** each task will require a Start predecessor and a Finish successor. Most schedules may be created using only Finish-to-Start relationships with positive or negative lags. This method ensures a Closed Network is created and the **Critical Path** flows through the tasks and not just through the relationships. When Start-to-Start relationships are used the Critical Path is calculated through the relationships only and is not a true Critical Path; the Critical Path must run through the task to be a true Critical Path.

A delay to the completion of the first task in the example on the left below will not delay subsequent tasks as the calculated Critical Path flows through the relationships and not the tasks; therefore a true Critical Path has not been created. The example on the right has created a true Critical Path because the driving path flows through the tasks and the relationships:

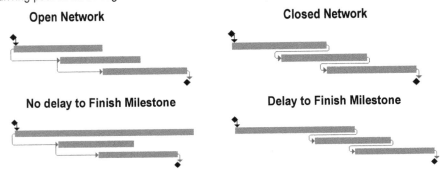

There is no simple method of documenting which is hard and which is soft logic. A schedule with a large amount of soft logic has the potential of becoming very difficult to maintain when the plan is changed. You will also find that soft logic converts to hard logic as a project progresses when commitments are made and tasks have started.

Constraints

Constraints are applied to tasks when relationships do not provide the required result. Typical applications of a constraint are:

- The availability of a site to commence work.
- The supply of information by a client.
- The required finish date of a project.

Constraints are often entered to represent contract dates or **External Logic** and may be directly related to contract items. It is often useful to make notes about contract dates reflected in the schedule in either the Task Note area or one of the Text columns. Constraints are covered in detail in the **CONSTRAINTS** chapter.

Two other terms you must understand are:

- **Predecessor**, a task that controls the start or finish of another immediate subsequent task.
- **Successor**, a task whose start or finish depends on the start or finish of another immediately preceding task.

In the picture above A is the predecessor of B and C is the successor of B.

There are four types of dependencies available in Microsoft Project:

- Finish-to-Start (**FS**) – this is the default and also known as conventional
- Start-to-Start (**SS**)
- Finish-to-Finish (**FF**)
- Start-to-Finish (**SF**) – this relationship type is rarely used and should be avoided because it places the successor before the predecessor. On the other hand it does work like a Zero Free Float constraint which is not available in Microsoft Project.

The following pictures show how the dependencies appear graphically in the **Gantt Chart** and **Network Diagram** (PERT) views:

The **FS** (or conventional) dependency looks like this:

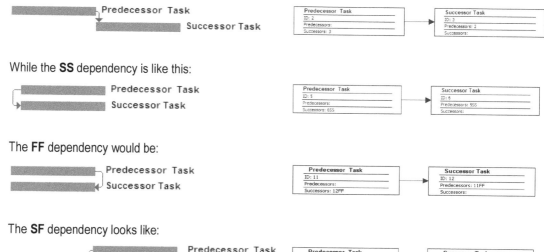

While the **SS** dependency is like this:

The **FF** dependency would be:

The **SF** dependency looks like:

 A **SF** relationship is often considered bad scheduling and not logical as the successor is before the predecessor.

Other software like Oracle Primavera and Asta Powerproject allow multiple relationships between two activities which allows far more flexibility.

Large negative lags are normally unacceptable and Ladder Scheduling is used to link a set of tasks that have substantial overlap, such as pipe laying operations. Most products allow multiple relationships between two tasks, as per the P6 example below, where the tasks are linked using two relationships, a SS+3d and a FF+3d:

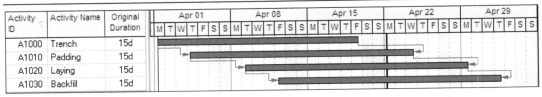

Microsoft Project does not allow two relationships between tasks. Ladder Scheduling may be achieved by:

- Commencing a chain of tasks with a Start Milestone,

- Connect the Start Milestone to each task with a Start to Start plus the appropriate lag and

- Connect each task to their successor with a Finish to Finish relationship plus the appropriate lag:

9.2   Understanding Leads and Lags

A **Lag** is a duration that is applied to a dependency to make the successor start or finish earlier or later and may be applied to any relationship type.

- A successor task will start or finish later when a positive **Lag** is assigned. Therefore, a task requiring a 3-day delay between the finish of one task and start of another will require a positive lag of 3 days.

- Conversely, a lag may be negative (also called a **Lead**) when a new task can be started before the predecessor task is finished.

- **Leads** and **Lags** may be applied to any relationship type including summary task relationships. Dependencies between summary tasks should be avoided and all relationships should be placed on detail tasks which makes following the logic simpler.

An example of an **FS** with positive lag:

An example of an **FS** with negative lag:

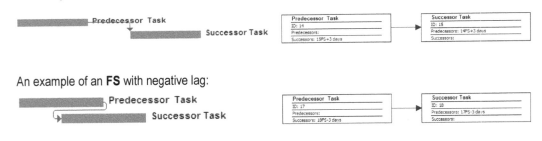

Here are some important points to understand about lags.

- Lags are calculated on the **Successor Calendar** except with Microsoft Project 2003 and earlier which uses the **Project Calendar**, set in the **Project Information** form. Therefore files with lags and multiple calendars may calculate differently in earlier versions. (Oracle Primavera SureTrak and P3 use the predecessor's calendar, Primavera P6 has four user options and Asta Powerproject puts the lag on either or both the predecessor or successor tasks and not the relationship.)

- Lags may be assigned **Elapsed** durations, therefore they will be based on a 24-hour, 7-day per week. To enter an elapsed lag type an "**e**" before the unit, e.g. **5 ed**.

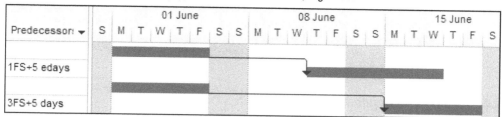

 The use of Finish to Start relationships with positive lags is often considered bad practice as it usually represents a missing activity.

 Other software like Asta Powerproject put the lead or lag on the activity and not the relationship, thus the lead or lag may be assigned the predecessor or successor calendar.

- Lags may be expressed in terms of percentage and in this situation the lag is a percentage of the predecessor's duration. The example below shows an FS dependency +250%; the predecessor is 1 day so the lag is 2.5 days. This results in the successor of 1 day duration spanning 2 days.

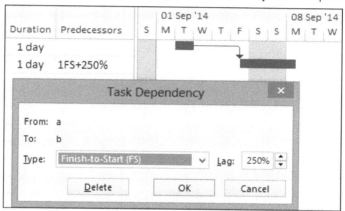

 You must be careful when using a lag to allow for delays such as curing concrete when the Successor Calendar is not a 7-day calendar. Since this type of task lapses nonworking days, the task could finish before Microsoft Project's calculated finish date. You may want to use elapsed durations in a lag in this situation.

9.3   *Restrictions on Summary Task Dependencies*

Dependencies may be made between summary tasks and detail tasks of a different summary task. Consider the following points when using dependencies at summary task level:

- There is a built-in dependency between summary and detail tasks. Detail tasks may be considered as Start-to-Start successors and Finish-to-Finish predecessors of their summary task.
- Summary tasks may only have **FS** and **SS** dependencies; you will receive a warning message when you attempt to enter an illegal dependency.

It is recommended that dependencies be maintained at the detail level. This is particularly important when moving tasks from one summary task to another since the dependencies will still be valid. Again, be aware that there is a function found under the **File**, **Options**, **Schedule** tab, **Scheduling options for this project:** section, **Autolink inserted or moved tasks** which will change relationships when tasks are moved. This function should be turned off if you wish to move a task and keep the existing logic.

9.4   Assigning Dependencies

The dependencies may be displayed or hidden with the **Layout** form.

- Select **Format**, **Format** group, **Layout** to open the **Layout** form and click on the radio button under the style you require:

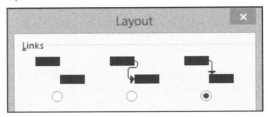

- The color of the dependency line is inherited from the color of the predecessor task.
- To display a Critical Path on the relationship lines you will need to format the bars as critical.

9.4.1   Graphically Adding a Dependency

You may graphically add a **Finish-to-Start** dependency only by:

- Selecting the **Gantt Chart**, **Calendar** or **Network Diagram**, then:

- Move the mouse pointer over a task until the mouse pointer changes to a ⌖, left-click and drag to the successor task. The cursor will change to a ◷ shape during this operation.

9.4.2   Using the Link and Unlink Buttons

The ⌗ **Link Tasks** button found the **Task**, **Schedule** group, which you should add to the **Quick Access Menu** may be used for linking tasks with a Finish-to-Start dependency:

- Highlight one or more tasks using Ctrl and left-click (to select one task at a time) or Shift and left-click (to select a contiguous group of tasks).

- Then click the ⌗ **Link Tasks** button on the toolbar and the tasks will be linked with Finish-to-Start dependencies in the order that they were selected.

To remove a dependency, select the tasks and click on the ⌗ **Unlink Tasks** button.

 The limit of a maximum of 10 groups of tasks that could be linked in this way has been removed from Microsoft Project 2010.

This function is not end sensitive as in Primavera products and Asta Powerproject and may only be used for assigning a Finish-to-Start relationship.

9.4.3 Task Linking Using the Keyboard

Multiple tasks may be linked with a Finish-to-Start dependency:

- Highlight one or more tasks using Ctrl-left-click (to select one task at a time) or Shift and left-click (to select a group of tasks).

- Then select **Ctrl+F2** and the tasks will be linked with Finish-to-Start dependencies in the order that they were selected.

- A group of tasks is created by dragging over two or more tasks. These are linked with FS relationships.

 Primavera P3 and SureTrak software link tasks that have been highlighted from top to bottom and not in the order they are selected. Oracle Primavera P6 links in the order they are highlighted. Microsoft Project also links the tasks in the order they are highlighted.

9.4.4 Relationship Listing Issue

In Microsoft Project 2013 and 2016 the predecessor and successor tasks are displayed in Alphabetical order, not in Task ID order.

Therefore when you have large projects with a number of tasks with the same name it becomes difficult to know what activity you are looking for in the list of tasks when assigning relationships, see the picture on the left below:

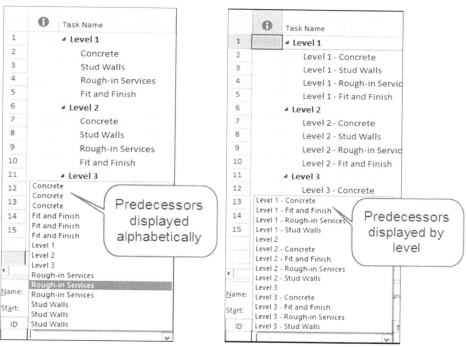

Consequently it is recommended that you consider making each description unique, see the picture above on the right.

The technique to do this with Excel was discussed in para Copying Tasks from Other Programs technique.

9.4.5 Adding and Deleting Predecessors with the Task Information Form

The **Task Information** form may be used for adding and deleting predecessors only.

- Double-click on a task to open the **Task Information** form,
- Select the **Predecessors** tab,
- To select the predecessor, you may either:
  - ➤ Type in the Predecessor Task ID in the first line under ID, or
  - ➤ Use the drop-down box under task name:

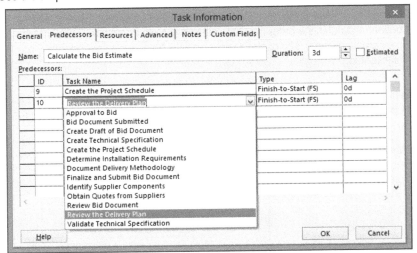

- Now enter the Relationship Type from the **Type** drop-down list and the lag, if required, from the **Lag** drop-down list.

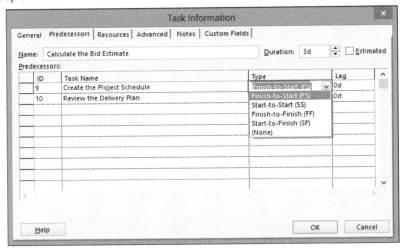

- To enter another relationship click on the next line.
- You are not able to scroll up or down to another task while the **Task Information** form is open.

To complete your operation, either:

- Press the **Enter Key** or click on the [OK] button to commit the changes, or
- Click on the **Esc Key** or click on the [Cancel] button to abort any changes.

9.4.6 Predecessor and Successor Details Forms

The Predecessor and Successor Details form may be displayed by:

- Opening the bottom pane by right-clicking in the Gantt Chart and selecting **Show Split**,

- Then making the bottom pane active by clicking anywhere in the bottom pane. The bar on the left-hand side of the bottom pane will turn blue when it is active,

- Displaying the **Task Details Form**, **Task Entry** or **Task Form** (shown below which is displayed by default),

- Then right-clicking in the bottom pane and select **Predecessors and Successors** to display the **Predecessors and Successors Detail** form:

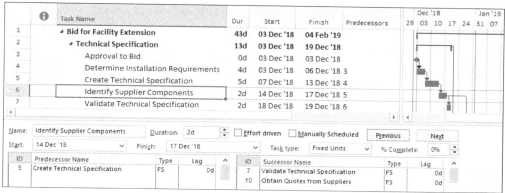

Predecessors and successors may be added using the same method as in the **Task Information** form.

 Double-clicking on a Predecessor or Successor in any of these forms will display the Predecessor or Successor **Task Information** form, allowing the dates, constraints, etc. of related tasks to be examined.

9.4.7 Editing or Deleting Dependencies Using the Task Dependency Form

To use the **Task Dependency** form, a logic link between tasks must already exist. To open the **Task Dependency** form, double-click on a task link (relationship line) in the Bar Chart or Network Diagram.

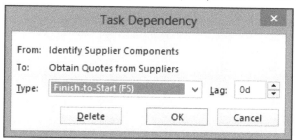

A link may only be edited or deleted from this form.

9.4.8 Autolink New Inserted Tasks or Moved Tasks

This function automatically creates predecessors to tasks above it and successors below it when a task is moved or inserted. This option may be activated by selecting the **File**, **Options**, **Schedule** tab, **Scheduling options for this project:** section and checking the **Autolink inserted or moved tasks** box.

When **Autolink inserted or moved tasks** is activated you must ensure that: you have selected the whole task by clicking on the Task ID and therefore highlighting all the columns before you drag the task to a new location. Otherwise, you will only move the cell contents.

 When the task is moved, **Autolink inserted or moved tasks** will change the existing predecessor and successor logic without warning. This function potentially makes substantial changes to your project logic and may affect the overall project duration. It is suggested that the option is **NEVER** switched on as dragging a task to a new location may completely change the logic of a schedule.

9.4.9 Editing Relationships Using the Predecessor or Successor Columns

The **Predecessor** or **Successor** column may be displayed and edited following the example below.

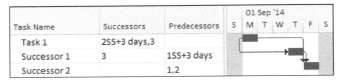

9.5 Scheduling the Project

Once you have your tasks and logic in place, Microsoft Project calculates the tasks' dates/times. More specifically, Microsoft Project has **Scheduled** the project to calculate the **Early Dates**, **Late Dates** and **Float**.

The calculation method is outline in detail in para 10.7 of the **NETWORK DIAGRAM VIEW** chapter.

This will allow you to review the **Critical Path** of the project. Microsoft Project uses the term **Slack** instead of the standard term **Float**. Both terms are used interchangeably throughout this book.

Sometimes it is preferable to prevent the **Automatic Calculation** of your project's start/end dates. To do this, select **File**, **Options**, **Schedule** tab, **Calculation**, click on **Off**. This is the same as **Manual Calculation** in earlier versions of the software.

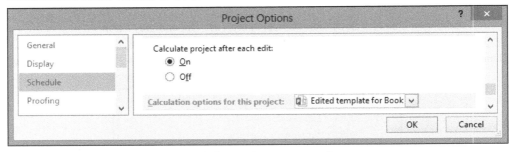

To calculate the schedule with the calculation mode set to manual:

- Press the **F9 Key**, or
- Click on the **Select All** button, top left-hand corner of the Gantt Chart view, right-click to open a menu and select ⊞ **Calculate Project**, or
- Add a **Schedule Project** button to the **Ribbon** or **Quick Access Toolbar**.

9.6 Workshop 7- Adding the Relationships

Background

You have determined the logical sequence of tasks, so you may now enter the relationships.

Assignment

1. Ensure you have the last workshop file open.

2. Apply the **Gant Chart Inc Total Float and Neg Float** view.

3. Apply the **Entry** table.

4. Remove the **Task Calendar** column.

5. Add the **Predecessor** and **Total Slack** columns if they are not displayed.

6. Input the logic below using several of the methods detailed in this chapter.

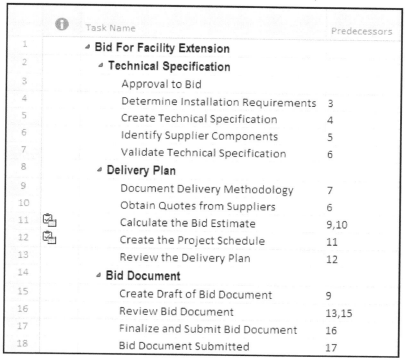

	Task Name	Predecessors
1	⊿ **Bid For Facility Extension**	
2	⊿ **Technical Specification**	
3	Approval to Bid	
4	Determine Installation Requirements	3
5	Create Technical Specification	4
6	Identify Supplier Components	5
7	Validate Technical Specification	6
8	⊿ **Delivery Plan**	
9	Document Delivery Methodology	7
10	Obtain Quotes from Suppliers	6
11	Calculate the Bid Estimate	9,10
12	Create the Project Schedule	11
13	Review the Delivery Plan	12
14	⊿ **Bid Document**	
15	Create Draft of Bid Document	9
16	Review Bid Document	13,15
17	Finalize and Submit Bid Document	16
18	Bid Document Submitted	17

7. Add the **Format**, **Format** group, 🖳 **Layout** button to the **Quick Access Toolbar** if it is not on the toolbar.

8. Hide and display the Logic Links using **Format**, **Format** group, **Layout,** to open the **Layout** form (If your links are displayed by default, hide and then display them again.)

9. Check your results against the pictures below:

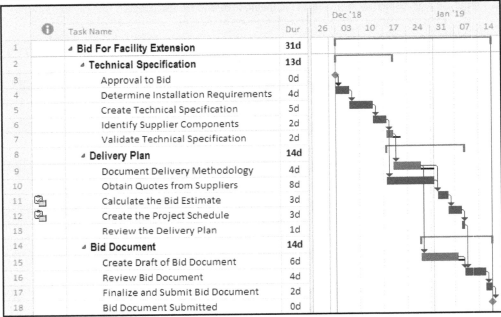

		Task Name	Dur	Start	Finish	Total Slack	Predecessors
1		⊿ Bid For Facility Extension	31d	03 Dec '18	17 Jan '19	0d	
2		⊿ Technical Specification	13d	03 Dec '18	19 Dec '18	0d	
3		Approval to Bid	0d	03 Dec '18	03 Dec '18	0d	
4		Determine Installation Rec	4d	03 Dec '18	06 Dec '18	0d	3
5		Create Technical Specifica	5d	07 Dec '18	13 Dec '18	0d	4
6		Identify Supplier Compone	2d	14 Dec '18	17 Dec '18	0d	5
7		Validate Technical Specific	2d	18 Dec '18	19 Dec '18	2d	6
8		⊿ Delivery Plan	14d	18 Dec '18	09 Jan '19	0d	
9		Document Delivery Metho	4d	20 Dec '18	27 Dec '18	2d	7
10		Obtain Quotes from Suppl	8d	18 Dec '18	31 Dec '18	0d	6
11	📇	Calculate the Bid Estimate	3d	02 Jan '19	04 Jan '19	0d	9,10
12	📇	Create the Project Schedul	3d	05 Jan '19	08 Jan '19	0d	11
13		Review the Delivery Plan	1d	09 Jan '19	09 Jan '19	0d	12
14		⊿ Bid Document	14d	28 Dec '18	17 Jan '19	0d	
15		Create Draft of Bid Docum	6d	28 Dec '18	07 Jan '19	2d	9
16		Review Bid Document	4d	10 Jan '19	15 Jan '19	0d	13,15
17		Finalize and Submit Bid Do	2d	16 Jan '19	17 Jan '19	0d	16
18		Bid Document Submitted	0d	17 Jan '19	17 Jan '19	0d	17

		Task Name	Dur
1		⊿ Bid For Facility Extension	31d
2		⊿ Technical Specification	13d
3		Approval to Bid	0d
4		Determine Installation Requirements	4d
5		Create Technical Specification	5d
6		Identify Supplier Components	2d
7		Validate Technical Specification	2d
8		⊿ Delivery Plan	14d
9		Document Delivery Methodology	4d
10		Obtain Quotes from Suppliers	8d
11	📇	Calculate the Bid Estimate	3d
12	📇	Create the Project Schedule	3d
13		Review the Delivery Plan	1d
14		⊿ Bid Document	14d
15		Create Draft of Bid Document	6d
16		Review Bid Document	4d
17		Finalize and Submit Bid Document	2d
18		Bid Document Submitted	0d

10. Save your **OzBuild Bid** project.

10 NETWORK DIAGRAM VIEW

The **Network Diagram View** is an enhancement of the earlier versions of Microsoft Project **PERT View** and displays tasks as boxes connected by the relationship lines. This chapter will not cover this subject in detail but will introduce the main features.

To view your project in the Network View:

- Select **View**, **Task Views** group **Network Diagram**, or
- Select **Task**, **View** group and select **Network Diagram** from the drop-down list.

Topic	Menu Command
• Add new tasks	Click and drag into a blank area. This will add a new task with a Finish-to-Start relationship, or Use the **Insert Key** and a new task is added after the highlighted task without a relationship.
• Delete tasks	Select the task/s and press the **Delete** key.
• Display the **Task Information** form	Double-click on a Task Box, or Highlight a task and right-click on it then select **Information…** for the menu.
• Format Dependencies	Select **Format**, **Format** group, **Layout**
• Display the **Task Dependency** form	Double-click on a relationship line.
• Display information in the bottom pane	Right-click and select **Show Split**, or Drag the dividing bar.
• Format Task Boxes	Select **Format**, **Format** group, **Box Styles**, or Double-click on the outside edge of a box.
• Format an individual Task Box	Select **Format**, **Format** group, **Box**
• Change the scale of the display	Select **View**, **Zoom** group, **Zoom**, or Use the **Zoom Slider** at the bottom right-hand side of the screen.

10.1   Understanding the Network Diagram View

The following list describes the main display features of the Network Diagram View:

- Summary tasks, detail tasks and Milestones are normally formatted with a different shape. Typical shape examples are provided as follows:

    ➢ Summary tasks are trapezoidal –

    ➢ Detail tasks are rectangular –

    ➢ The Milestone Task is an elongated diamond –

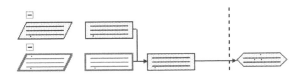

- Summary tasks are positioned to the left at the same level or above detail tasks.
- Summary tasks may be rolled up by clicking on the ⊟ above the task and expanded by clicking on the ⊞ above a rolled up summary task.
- All relationship lines are drawn as **Finish-to-Start** even when they are not linked as a **Finish-to-Start** relationship. The link type may be displayed on the arrow. See **Link style** later in this chapter.
- Critical tasks may be formatted to have different borders and backgrounds.
- The contents of the Task Boxes and relationship lines may also be formatted.

10.2   Adding and Deleting Tasks in the Network Diagram View

Adding and deleting Tasks in the Network Diagram View:

- A **New Task** may be created with a **Finish-to-Start** relationship by dragging from the center of a Task Box into a blank part of the screen.
- A **New Task** may be created without a relationship below the tasks position in the Gantt Chart by:

    ➢ Using the **Insert Key**, or

    ➢ Selecting **Task**, **Insert** group, **Task** ⊞ **Task** (New Task).

- Double click on the task box and this will open the Task Information form enabling you to edit the **Task Name**, **Duration**, **Calendar** or any other attribute of a new task available in the **Task Information** form.

10.3 Adding, Editing and Deleting Dependencies

Dependencies may be added, edited or deleted using the following methods:

- Graphically add a relationship by clicking on the center of one task and dragging to the successor.

- Hold the **Ctrl Key** to select two or more tasks and then use the **Task**, **Schedule** group, ⊂∞ **Link Task** function.

- The **Task**, **Schedule** group, ⊊ **Unlink Task** function removes dependencies between selected tasks.

- Open the **Task Information** form by double-clicking on a task and selecting the **Predecessors** tab.

- Double-click on a **Relationship line** to open the **Task Dependency** form and edit or delete a dependency.

- Create a split window by right-clicking and selecting **Show Split** and then display the predecessors and successors in the bottom pane by right-clicking and selecting **Predecessors and Successors**.

10.4 Formatting the Task Boxes

Task Boxes may be formatted from the **Box Styles** form, which is displayed by:

- Selecting **Format**, **Format** group, ⟶ **Box Styles**:

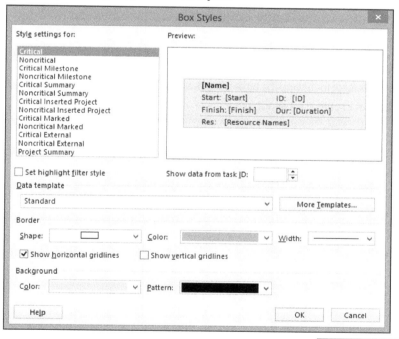

- A new template may be created to display other data by clicking on the ⬚ More Templates... button which will open the **Data Templates** form.

10.5   Formatting Individual Boxes

Once highlighted, a **Task Box** may be formatted differently from all the others by:

- Selecting **Format**, **Format** group, 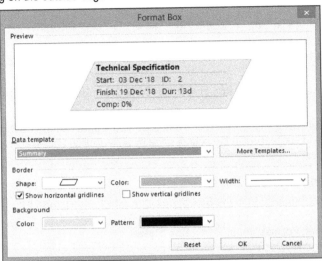 **Box** to open the **Format Box** form, or

- Double-clicking on the outside edge of a box.

The ☐ Reset ☐ button is used to set the formatting back to default.

10.6   Formatting the Display and Relationship Lines

Most formatting, except formatting the boxes, is set within the **Layout** form. Select **Format**, **Format** group, 🔲 **Layout** to open the **Layout** form:

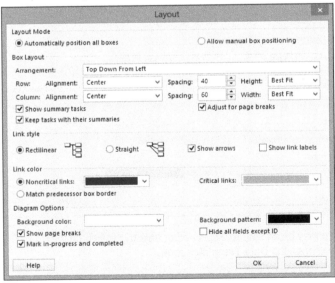

- The **Layout Mode** allows you to drag a Task Box to a new position. When **Allow manual box positioning** is selected, place the mouse over the Task Box and then when it changes to a ✛, drag the box to the required location.

- **Box Layout** allows you to specify how the Task Boxes are arranged. This option will only work when **Layout Mode** is set to **Automatically position all boxes**. For example, when "Top Down by Week" is selected, all the Tasks that start in the first week will be placed in the first column. The other options under **Box Layout** should be self-explanatory.

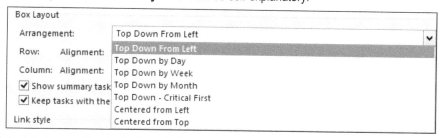

- **Link style** is an option to display relationships. Checking **Show link labels** will place the relationship type and lag on the relationship line and is useful when all your relationships are not Finish-to-Start. See below.

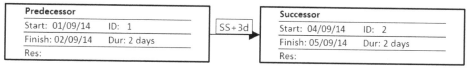

- **Link color** allows you to specify the color of the critical links and other link types. Selecting the option **Match predecessor box border** sets the format of the relationship lines to the predecessor's box border format. The formatting of links is not available in the Gantt Chart view.

- **Diagram Options** are self-explanatory. The **Hide all fields except ID** will display the Task Boxes, as below. In the example, note that **Show link labels** has also been checked:

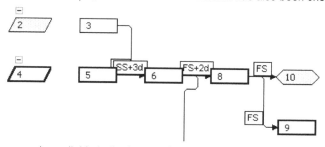

- Many of the commands available in the **Layout** form are also available on the **Ribbon** toolbar, **Format** tab:

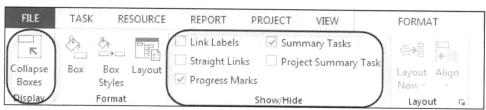

10.7 Early Date, Late Date and Float/Slack Calculations

To help understand the calculation Early Dates, Late Dates, Total Float and Critical Path, we will now manually work through an example. The boxes below represent tasks on a 7 day per week calendar.

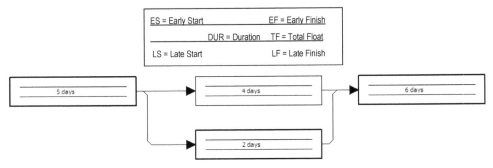

The forward pass calculates the early dates: EF = ES + DUR − 1

Start the calculation from the first task and work forward in time.

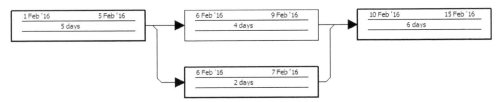

The backward pass calculates the late dates: LS = LF − DUR + 1

Start the calculation at the last task and work backward in time.

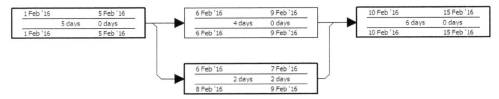

The **Critical Path** is the path where any delay causes a delay in the project end date and runs through the top row of tasks.

Total Float is the difference between either the **Late Finish** and the **Early Finish** or the difference between the **Late Start** and the **Early Start**.

- The 2 days' task has float of 9 − 7 = 2 or 8 − 6 = 2 days. None of the other tasks has float.

The example above would look like the picture below in the Gantt Chart view:

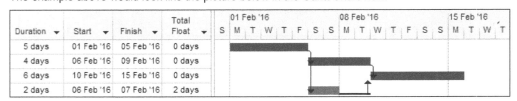

10.8 *Workshop 8 - Scheduling Calculations*

Background

We want to look at the Network Diagram and practice calculating Early and Late dates with a simple manual exercise.

Assignment

1. Apply the **Network Diagram** view by right clicking in the dark band on the left hand side of the screen and selecting it.

2. Hide the Summary tasks by selecting **Format**, **Show/Hide** group, **Summary Tasks**.

3. Use the 🔍 🔍 buttons to zoom in and out.

4. Right-click in the **Network Diagram,** select **L**ayout to open the **Layout** form and select **Allow manual box positioning**.

5. You should now be able to reposition the boxes manually by left-clicking and dragging.

6. Reapply the **Gantt Chart** View.

7. Calculate the Early Dates, Late Dates, and Total Float for the following activities, assuming a Monday-to-Friday working week and the first activity starting on 01 Apr 19.

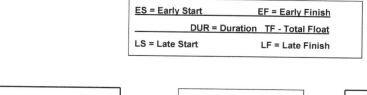

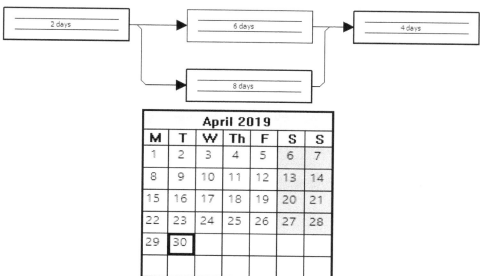

8. See over the page for the answer:

Answer to Workshop 8

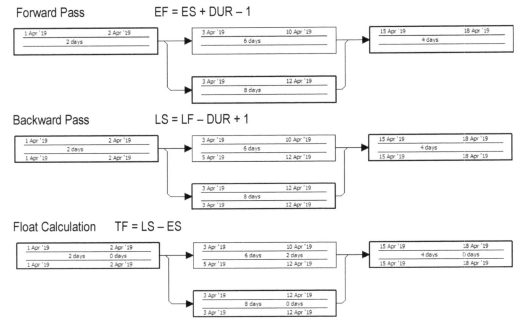

9. The Early Bar is the upper bar, the Late Bar the lower bar and the end of the Total Float bar, which is the thin bar, ends at the Late Finish date.

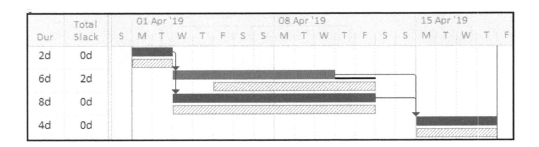

11 CONSTRAINTS

Constraints are used to impose logic on tasks that may not be realistically scheduled with logic links. Microsoft Project will only allow one constraint against a given task with the exception of a **Deadline Date**. This chapter will deal with the following constraints in detail, which are the minimum number of constraints that are required to effectively schedule a project:

- **Start No Earlier Than**
- **Finish No Later Than**

These two constraints may be applied to summary and detail tasks; all others may only be applied to detail tasks.

- **Start No Earlier Than** (also known as an "Early Start" constraint) is used when the start date of a task has been set by a client or an external event. Microsoft Project will not schedule the task early start date prior to this date.

- **Finish No Later Than** (also known as a "Late Finish" constraint) is used when the latest finish date is stipulated. Microsoft Project will not show the task's Late Finish date after this date, but the Early Finish date will be able to exceed this date and Negative Float will be generated when the Late Date is earlier than the Early Date. This represents the amount of time to be "caught up" to finish the project on time.

The following chart summarizes the methods used to assign Constraints to Tasks:

Topic	Notes for Creating a Constraint
• Open the **Task Information** form.	Double-click on the task and select the **Advanced** tab.
• Display the **Constraint Type** and **Constraint Date** columns.	Apply the **Constraint Table** or insert the columns in an existing table.
• Type a date into a task Start or Finish date field of the **Task Information** form or the **Details** form or a column.	A date typed in a task Start field will apply a **Start No Earlier Than** constraint of that date, without warning. A date typed in a task Finish field will apply a **Finish No Later Than** constraint, without warning.
• Display a **Combination** view displaying the **Task Details** form in the bottom pane.	Open the bottom pane by right-clicking in the Gantt Chart and selecting **Show Split**, Right-click on the vertical band on the bottom left-hand side of the screen labeled **Task Form** and select from the menu **Task Details** form.

There is an option found under the **File**, **Options**, **Schedule** tab, **Scheduling options for this project:** section, titled **Tasks will always honor their constraint dates**. This function is described in detail in the **OPTIONS** chapter, **Schedule** section. When this option is checked, which is the default, constraints override a logic link. When unchecked, a logic link will override a constraint and a task may be delayed.

Tasks will always honor their constraint dates should never be checked.

Only one constraint may be applied to a task or milestone except when a **Deadline Date** is assigned. Oracle Primavera products allow two constraints per task and Asta Powerproject has a Schedule Between constraint which is effectively assigning two constraints to a task.

A full list of **constraints** available in Microsoft Project is:

- **As Soon As Possible**   This is the default for a new task. A task is scheduled to occur as soon as possible and does not have a Constraint Date.

- **As Late As Possible**   A task will be scheduled to occur as late as possible and does not have any particular Constraint Date. The Early and Late dates have the same date. A task with this constraint has no Total Float and will delay the start of all the successor tasks that have Total Float.

- **Start No Earlier Than**   This constraint sets a date before which the task will not start.

- **Start No Later Than**   This constraint sets a date after which the task will not start.

- **Must Start On**   This constraint sets a date on which the task will start. Therefore the task has no float. The early start and the late start dates are set to be the same as the Constraint Date.

- **Must Finish On**   This constraint sets a date on which the task will finish. Therefore has no float. The early finish and the late finish dates are set to be the same as the Constraint Date.

- **Finish No Earlier Than**   This sets a date before which the task will not finish.

- **Finish No Later Than**   This sets a date after which the task will not finish.

- **Deadline Date**   This is similar to applying a **Finish No Later Than** constraint. This offers the opportunity of putting a second constraint on a task.

The author recommends that constraints should be used as sparingly as possible and only used to represent events that may not be reasonably represented with relationships. These would often be events outside the control of the project manager such as the date a client will release a site to commence work or a predetermined time a project must be complete by.

Earlier Than constraints operate on the **Early Dates** and **Later Than** constraints operate on **Late Dates**. The picture below demonstrates how constraints calculate Total Float (Total Slack) of tasks (without predecessors or successors) against the first task of 10 days' duration:

Constraint Type	Constraint Date	Early Start	Early Finish	Late Start	Late Finish	Total Slack
As Soon As Possible	NA	15 Nov	28 Nov	15 Nov	28 Nov	0d
As Late As Possible	NA	15 Nov	17 Nov	24 Nov	28 Nov	0d
Start No Earlier Than	21 Nov	21 Nov	23 Nov	24 Nov	28 Nov	3d
Start No Later Than	22 Nov	15 Nov	17 Nov	22 Nov	24 Nov	5d
Must Start On	22 Nov	22 Nov	24 Nov	22 Nov	24 Nov	0d
Must Finish On	18 Nov	16 Nov	18 Nov	16 Nov	18 Nov	0d
Start No Earlier Than	22 Nov	22 Nov	24 Nov	24 Nov	28 Nov	2d
Finish No Later Than	22 Nov	15 Nov	17 Nov	18 Nov	22 Nov	3d

1. An **Expected Finish** is used in Oracle Primavera software to calculate the remaining duration of a task. There is no equivalent of this constraint in Microsoft Project.

2. A task assigned with an **As Late as Possible** constraint in Oracle Primavera software will schedule the task so it absorbs only **Free Float** and will not delay the start of successor tasks. In Microsoft Project, a task assigned with an **As Late as Possible** constraint will be delayed to absorb the Total Float and delay all its successor tasks, not just the task with the constraint. Therefore an **As Late as Possible** constraint should be used with care in Microsoft Project.

3. The **Must Start On** and **Must Finish On** are hard constraints and will not allow Float to calculate through the constraint dates and will effectively cut the schedule into two.

11.1 Assigning Constraints

11.1.1 Using the Task Information Form

To assign a constraint using the **Task Information** form:

- Double-click on a task to open the form,
- Select the **Advanced** tab,
- Select the **Constraint type:** from the drop-down box,
- Select the **Constraint date:** from the calendar, or type the date in the box, and
- Click on the [OK] button to accept the constraint.

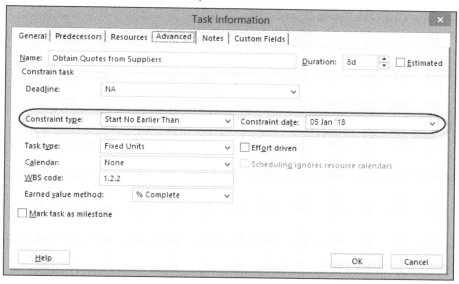

11.1.2 Using the Constraint Type and Constraint Date Column

To assign a constraint using the **Constraint Type** and **Constraint Date** columns:

- Either
  - ➢ Insert the **Constraint Type** and **Constraint Date** columns in your existing table by right-clicking on a column header and selecting **Insert Column**, or
  - ➢ Apply the **Constraint Dates** table by selecting **View**, **Data** group, **Tables**, **More Tables....**
- Assign your **Constraint Type** and **Constraint Date** from the drop-down boxes in the columns.

	Task Name	Duration	Constraint Type	Constraint Date
10	Obtain Quotes from Suppl	8d	Start No Earlier Than	03 Jan
11	Calculate the Bid Estimate	3d	As Soon As Possible	NA
12	Create the Project Schedul	3d	As Soon As Possible	NA
13	Review the Delivery Plan	1d	As Soon As Possible	NA
14	◢ Bid Document	14d	As Soon As Possible	NA
15	Create Draft of Bid Docum	6d	Finish No Later Than	26 Jan

11.1.3 Typing a Date into the Task Information or Details Form

You may assign some constraints from the **Task Information** or **Task Details** form:

- A **Start No Earlier** constraint is assigned by overtyping the Start date, and
- A **Finish No Earlier** constraint is assigned by overtyping the Finish date.

You may overtype or select a date from the drop-down box.

 This function provides no warning that a constraint has been set and new users need to be careful when they enter or copy and paste a date into a Start or Finish field.

11.1.4 Using the Task Details Form

Constraints may be set in the **Task Details** form, which is covered in detail in the **VIEWS AND DETAILS** chapter.

The **Task Details** form should first be made visible in the **View** menu so it may be more easily opened at other times by:

- Opening the **More Views** form available from many menus,
- Select the **Task Details Form**,
- Open the **Task Details Form - View Definition** by clicking on the [Edit...] button,
- Checking the **Show in menu** in the form.

Then **Task Details** form may be displayed by:

- Opening the bottom pane by right-clicking and selecting **Show Split**,
- Right-clicking on the vertical band on the bottom left-hand side of the screen labelled **Task Form,**
- Select from the menu the **Task Details** form:

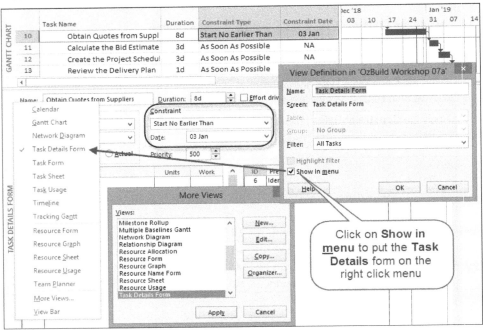

11.2   Deadline Date

Deadline Date was a new to Microsoft Project 2000 feature, which allows the setting of a date that a task should be complete. A **Deadline Date** is similar to placing a **Finish No Later Than** constraint on a task and affects the calculation of the **Late Finish** date and float of the task. A second constraint such as an Early Start constraint may also be assigned to a task with a Deadline Date.

The Deadline Date may be displayed as a column and the Indicator displayed on the bar chart may be formatted in the **Bars** form. This is covered in the **FORMATTING THE DISPLAY** chapter.

An Indicator icon is placed in the Indicator column when the Deadline Date creates Negative Float.

The **Move Project** function moves a Deadline Date, see para 16.4.1.

11.3   Changing Manually Scheduled Tasks to Auto Scheduled

When a task is changed from **Manually Scheduled** to **Auto Scheduled** in the **File**, **Options**, **Schedule** tab, **Scheduling options for this project:** section, **Keep Tasks on nearest working day when changing to Automatically Scheduled mode** will:

- When checked set a **Start No Earlier** constraint on the task so it does not move back in time if it has no predecessors, or

- When unchecked the task will not have a constraint set and will move back when it has no predecessors.

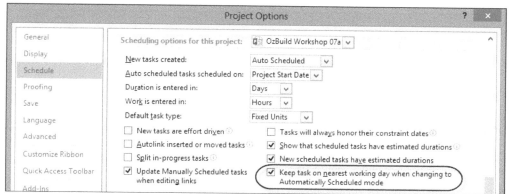

If your schedule is behaving strangely or there is unfamiliar bar formatting, this could be because some tasks have become **Manually Scheduled**. To fix this you may either:

- Display the **Task Mode** column and search for **Manually Scheduled** tasks and fix them in the column, or

- Click on the **Select All** button and then click on the Auto Schedule button.

11.4   Task Notes

It is often important to note why constraints have been set. Microsoft Project has functions that enable you to note information associated with a task, including the reasons associated for establishing a constraint.

The **Task Information** form has a **Note** tab, which has some word processing-type formatting functions.

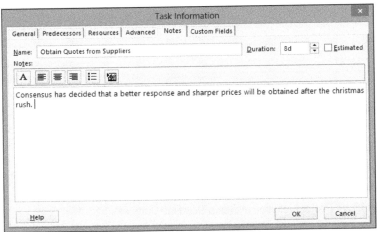

The notes may be displayed by:

- Inserting the **Notes** column. The Indicators column has a ✎ icon to indicate a note and a ⊞ icon to indicate a constraint:

	ⓘ	Task Name	Constraint Type	Constraint Date	Notes
10	⊞✎	Obtain Quotes from Suppliers	Start No Earlier Than	03 Jan	Consensus has decided that a better response and sharper prices will be obtained after the christmas rush.

- Displaying the note next to the task bar chart:

- There is an option for printing task notes in the **File, Print Settings** form and this option prints the Task ID, Task Name and Note on a separate sheet:

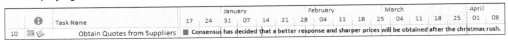

Other options for recording task notes are:

- Display and use one of the Text columns where the title may be renamed or customized.
- Insert a Text Box on the Gantt Chart using **Format**, **Drawing** group, **Drawings** and clicking on the Text Box, ⊡, and placing a note in the Text Box.

11.5 Completed Schedule Check List

At this point in time the schedule is complete and the following list could be used to check a completed schedule before submission:

- Check the full scope of the project has been captured.

- Ensure all contract or customer requirements have been considered.

- Check all calendars and holidays are correct and that the holidays are set well into the future so any delayed task will calculate correctly.

- Ensure there is a closed network with all tasks having a start predecessor and finish successor. This may be achieved by displaying the predecessor and successor columns.

- Check all relationships are valid.

- Check all constraints are valid and cross referenced to contract documentation or project management plan.

- Review float. Tasks with excessive float should be assigned dummy successors or delayed if they are not scheduled in a realistic timeframe with sequencing logic or Early Start constraints.

- Ensure all stakeholders are represented and agree to their scope or work.

- Ensure all risk mitigation tasks have been added.

- Check the Critical Path is realistic and aligned with what project personnel consider critical.

- A resourced schedule should be optimized by checking the histograms and tables.

- Evaluate the contingent time.

- Ensure all project personnel are in agreement with the schedule.

11.6 Workshop 9 - Constraints

Background

Management has provided further input to your schedule. The client requires the submission 28 Jan 19.

Assignment

1. Apply the **Gant Chart Inc Total Float and Neg Float** view and adjust the timescale to weeks and months.

2. Apply a **Finish No Later Than** constraint with a constraint date of 28 Jan 19 to task **18 Bid Document Submitted** task. If you are presented with an error message, read the message carefully and then set the constraint. Now review float, there should be no change in the Total Float. This demonstrates that a **Finish No Later Than** constraint after the calculated early finish date does not create positive float.

3. Due to proximity to Christmas, management has requested we delay task **10 Obtain Quotes from Suppliers** until first thing in the New Year, 03 Jan 19. It is hoped that sharper prices will be obtained after the Christmas rush. You may record this decision in the task notes.

   ➤ To achieve this, set a **Start No Earlier Than** constraint and a constraint date of 03 Jan 19 on task **10 Obtain Quotes from Suppliers**. Should you be presented with an error message, allow scheduling conflict and set the constraint.

   ➤ Now observe the impact on the Critical Path and end dates.

   ➤ Tasks 11 and 12 have 3 days Negative Float as they are assigned a 6 Day Week calendar:

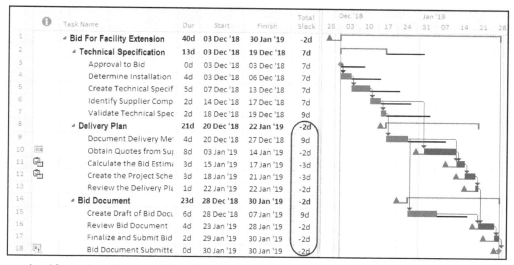

4. After review, it is agreed that 2 days can be deducted from task **16 Review Bid Document**. Change the duration of this task to 2 days.

5. Press **F9** to ensure the schedule is recalculated as Microsoft Project sometimes does not calculate the Float automatically.

6. You will notice that the Total Float of all critical tasks is now zero and the Critical Path runs between the two tasks with constraints.

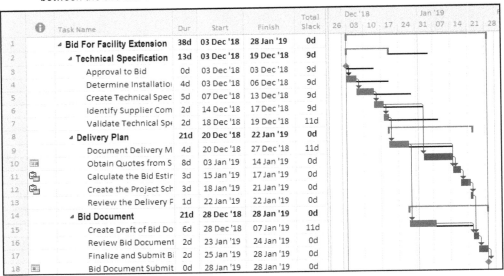

7. Save your **OzBuild Bid** project

8. Notice that activities with constraints have an icon in the indicators column.

9. **NOTE**: If your Total Float is not calculating as above, press **F9** to recalculate.

12 FILTERS

This chapter covers the use of **Filters** to select which tasks are displayed on the screen and in printouts.

12.1 Understanding Filters

Microsoft Project has the ability to display tasks that meet specific criteria. You may want to see only the incomplete tasks, or the work scheduled for the next couple of months, or the tasks that are in-progress, assigned a specific resource, responsibility, phase, discipline, system or belong to a physical area of a project.

Microsoft Project defaults to displaying all tasks. It has a number of predefined filters available that you may use or edit and you may also create one or more of your own.

There are **Task** filters that apply to **Task** views and **Resource** filters that apply to **Resource** views. Both types are created and applied in the same way. In this chapter we will focus on Task Filters but resource Filters are created in a similar way.

There are two types of filters:

- The first is where you select a **Filter** which exists or has been created using the **Filters** form.

- The second is to create an **AutoFilter** which is very similar to the Excel **AutoFilter** (Drop-down filter) function which is based on the selected column.

Topic	Menu Command
• Apply a **Filter**	Select **View**, **Data** group, **Filter:** drop-down box, and then either: • Select a filter from the list, or • Select **More Filters...** for a complete list of filters.
• Create or modify a **Filter**	Select **View**, **Data** group, **Filter:** drop-down box, ▽ **More Filters...** to open the **More Filters** form.
• Turn on **AutoFilter**	Select **View**, **Data** group, **Filter:** drop-down box, ▽ **Display AutoFilter**.
• Apply an **AutoFilter**	Click on the ▼ button in the column headers.

12.2 Understanding the Filter Menu

The **Filter** menu is opened from the **View**, **Data** group, **Filter:** drop-down box or you may put a **Filter** menu item on your **Quick Access Toolbar**. This may not display a complete list of available filters, as this list displays only filters that have been selected to be displayed in menu from the **Filter Definition** form and more may be available when selecting the **More Filters…** option.

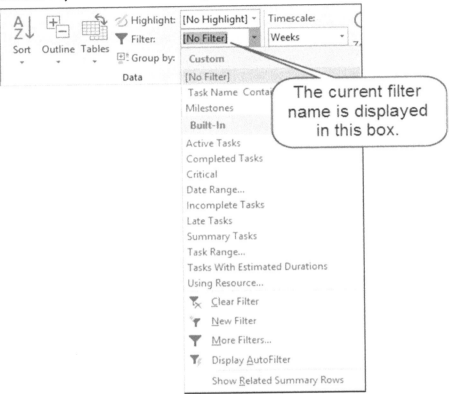

- **Custom** filters are those that the user has created or were created as part of a template.

- **Built-In** filters are those that exist as part of the Microsoft Project Global.mpt which is used to create a project when no template has been used.

- **Clear Filter** will remove all filters, both **Filters** and **AutoFilters**, and display all tasks.

- **New Filter** will open the **Filter Definition** form, allowing the creation of a new filter.

- **More Filters…** opens the **More Filters** form which will list all the filters available to the current project. Only filters with the **Show in menu** check box checked in the **Filter Definition** form are displayed in the menu above.

- **Display AutoFilter** turns on and off the ability to use AutoFilters and hides or displays the arrow in the column headers which in turn enables or disables the **AutoFilter** function:

- **Show Related Summary Rows** will display any related and hidden Summary Rows as a result of applying a filter.

12.3   Applying an Existing Filter

To apply a filter:

- Select **View**, **Data** group, **Filter:** and select a filter from the drop-down list, this lists only those filters selected to be shown in the menu.

- Select **View**, **Data** group, **Filter**, **More Filters…** to open the **More Filters** form to display all the filters that are available.

The **More Filters** form displays all filters available in a project; some of these may not be displayed in the Filter menu

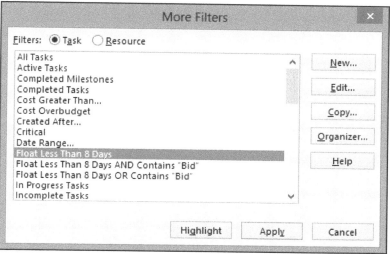

- The two radio buttons at the top of the form allow you to select filters that operate either on a **Task** view or on a **Resource** view.
  - ➢ **Task** filters will operate using most task data such as dates, durations, text columns, number columns, outline and WBS in Task Views.
  - ➢ **Resource** filters will operate on resource criteria only and not the task criteria in Resource Views.

- Select the required filter from the list, and

- Select either:
  - ➢ Highlight to shows all tasks but highlights the tasks that meet the criteria, or
  - ➢ Apply to display only the tasks that meet the criteria.

12.4 Creating and Modifying Filters

New filters are created from the **More Filters** form. This may be opened by selecting **View**, **Data** group, **Filters:**, **More Filters ...**:

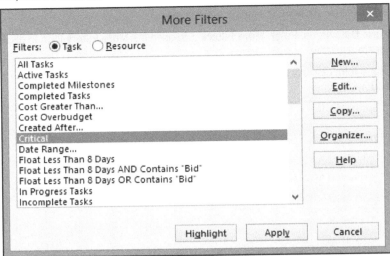

- Click on 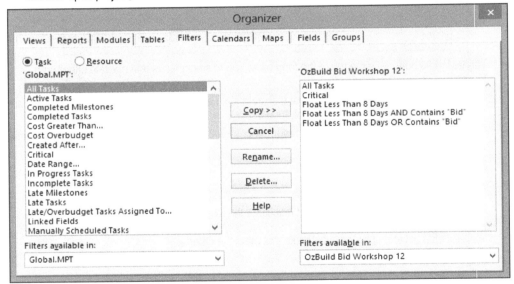 Organizer... to open the **Organizer** form where you may copy a filter to and from another open project, rename or delete a filter:

- Click on:
    - ➢ New... to create a new filter, or
    - ➢ Edit... to edit an existing filter, or
    - ➢ Copy... to copy an existing filter.

- The **Filter Definition** form will be displayed. The one-line example below will only display critical tasks.

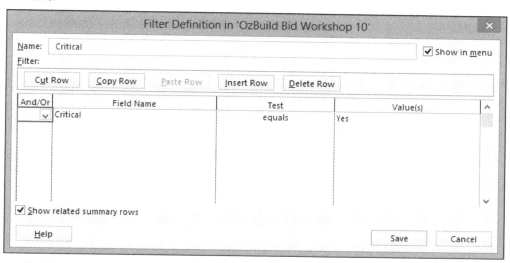

- The filter may be edited in the **Filter Definition** form to display or highlight the required tasks.
- Checking the **Show in menu** box will place the **Filter** in the menu.
- Checking **Show related summary rows** will display any associated Summary tasks.
- Select [Save] to return to the **More Filters** form.

 It is sometimes useful to place a space at the start of the filter name when creating a new filter as this will place the filter you have just created at the top of the **More Filters** form list.

12.5 Defining Filter Criteria

The filter criteria are determined by four columns of information in the **Filter Definition** form:

- **And/Or** option functions when you have two or more lines of data in the **Filter Definition** form to operate on.
- **Field Name** defines the data field you want to operate on.
- **Test** sets the criteria such as "Greater Than" or "Less Than" or "Equals."
- **Value(s)** is/are a date, number, Yes/No or text for the **Test** field to operate on. If more than one value is to be considered, for example when a **Test** is "Between," the two values are separated by a comma ",". The example below is a filter that will only display detail tasks that will start between 1 Jan '19 and 20 Jan '19.

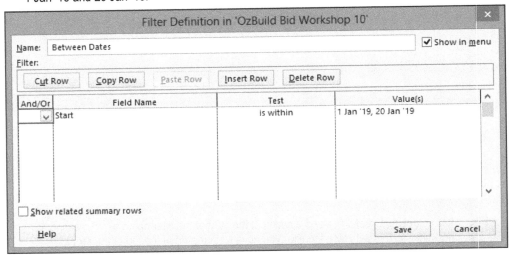

This chapter will not cover all the aspects of filter definitions but will cover the major principles, so you may experiment when you require a filter.

There are a number of predefined filters with a standard Microsoft Project installation. You should inspect these to gain an understanding of how filters are constructed and applied.

12.5.1 Simple Filters, Operator and Wild Cards

A simple **Filter** contains one line of data, therefore the **And/Or** function is not used. These are the most common filters and the most common filtering requirements.

These filters are used for purposes such as displaying tasks which:

- Are not started, complete, or in-progress tasks, or
- Have a Start or Finish before or after a particular date, or
- Contain specific text, or
- Are within a range of dates.

There are some operands you should be aware of:

- Some fields may have "**Yes**" or "**No**" entered in the **Value(s).** The task fields that may be filtered using a "**Yes**" or "**No**" include:
  - ➢ Summary – Indicates if the task is a summary or detail task,
  - ➢ Critical – Indicates if the task is critical when the Total Slack is less than or equal to the value entered in the **Calculation** tab or the **Options** form,
  - ➢ Milestones and
  - ➢ Effort-Driven.

  The filter below will select all tasks that are Milestones:

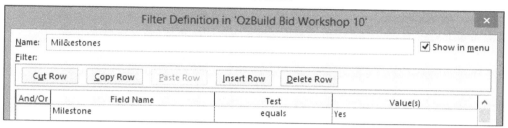

- The **Wildcard** functions are similar to the DOS Wildcard functions and are mainly used for filtering text:
  - ➢ You may replace a single character with a "**?**". Thus, a filter searching for a Task Name only containing "b**?**t" will display words like "bat", "bit" and "but".
  - ➢ The function is not case sensitive.
  - ➢ You may replace a group of characters with an *. Thus, a filter searching for a word in a Task Name containing "b*t" will display words like "blot", "blight" and "but".

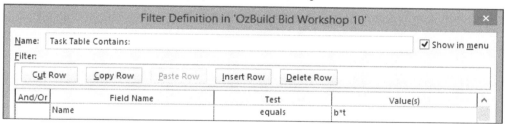

NOTE: For the Wildcard function to operate the **equals Test** must be used. This function does not work with other operands and in this mode works as a "**contains**" operand would operate.

- **NA** allows the selection of a blank value. The filter below displays tasks without either a Baseline Start or Baseline Finish date:

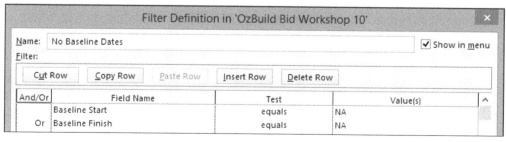

- A **Calculated Filter** compares one value with another. The **Value(s)** field is selected from the drop-down box. The example below will display those tasks that are scheduled to start later than the Baseline start:

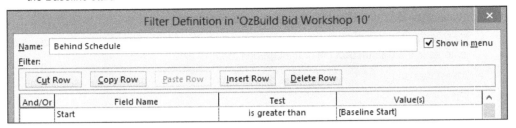

12.5.2 And/Or Filters

The **And/Or** function allows a search for tasks which meet more than one criterion by using the **And/Or** option:

- The filter below displays tasks with Slack less than 8 days **And** contain the word "Bid":

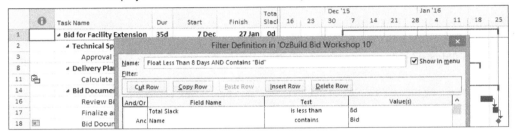

- The filter below is similar to the one above but displays tasks with Slack less than 8 days **Or** contain the word "Bid". It displays many more tasks:

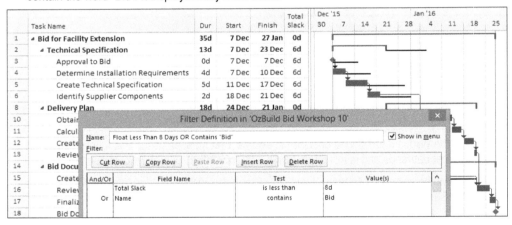

12.5.3   Multiple And/Or

Multiple **And/Or** statements are possible by placing a line with only an **And/Or** statement. The filter below selects tasks that have a Baseline Finish and are also scheduled to finish late or are not completing their work quickly enough, shown as the Budgeted Cost of Work Performed (Earned Value) is less than the Budgeted Cost of Work Scheduled (Planned Value):

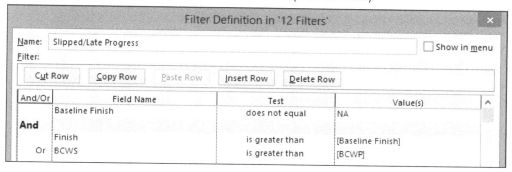

12.5.4   Interactive Filter

These filters allow you to enter the **Value(s)** of the filtered field after applying the filter. The filter is tailored each time it is applied via a user-prompt. The filter below will ask you to enter a word in the task name.

For this function to operate properly, the text in the **Value(s)** field must commence with a double quotation mark **"** and end with a double quotation and question mark **"?** :

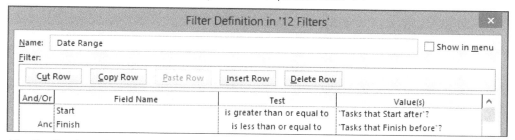

After the filter is applied, you will be presented with the **Interactive Filter** form to enter the required text:

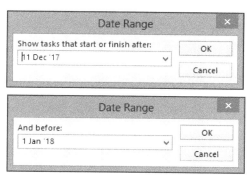

12.6 AutoFilters

This function was greatly enhanced in Microsoft Project 2010. Microsoft Project **AutoFilters** are similar to the Excel **AutoFilter** function and allow you to select the filter criteria from drop-down menus in the column headers. This function allows both **Grouping** and **Sorting** of tasks, which will be covered in the **TABLES AND GROUPING TASKS** chapter.

To create an **AutoFilter** based on one parameter:

- Turn on the **AutoFilter** function by selecting **View**, **Data** group, **Filter:** drop-down box, Display Auto**F**ilter.

- The column headers will display the 🔽 button in the column header. Click on this button in one of the columns to display a drop-down box:

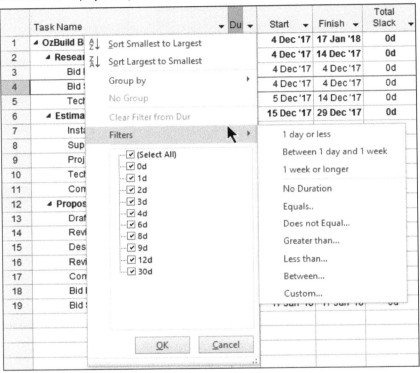

- Select the required criteria from the check box list, or
- Click on the **Filters** heading, as displayed above, to open up a second set of selections, which in turn will provide many more options and the **Custom...** will open the **Custom AutoFilter** form:

- Select the parameters you want to operate on from the four drop-down boxes and check the **And** or **Or** radio button.

- To save a drop-down filter as a normal filter, click on the ⬛ _Save..._ button, which opens the **Filter Definition** form.

- You may now select another column and create a filter based on two parameters to reduce the number of tasks displayed.

- After an AutoFilter has been applied the indicator in the column heading changes to a

  ⬛ Start ▽ .

 AutoFilters always select the associated summary tasks of any selected detail tasks. If you wish not to see Summary Tasks you should use a normal filter and not an Auto Filter.

12.7 Adding New Filters to the Global

Ensure the **File**, **Options**, **Advanced**, **Display**, **Automatically add new views, tables, filters, and groups to the global** is not checked.

When checked every new view, table or filter will be added to the Global.mpt and each time you create a new project it will contain more and more views, tables and filters:

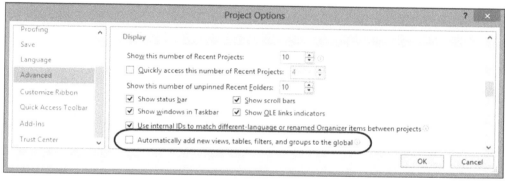

12.8 Copy a Filter to and from Another Open Project

Click on **Organizer** button in the **More Filters** form to open the **Organizer** form where you may copy a filter to and from another open project or the Global.mpt, rename or delete a filter:

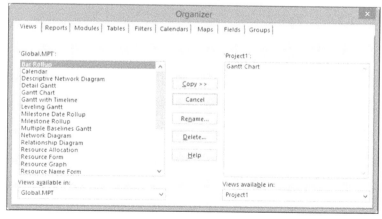

12.9   Workshop 10 - Filters

Background

Management has asked for some reports to suit their unique requirements.

Assignment

1. They would like to see all the critical tasks.

    ➢ Find the 🔽 **Filter** button, found under **View**, **Data** group, or in the **Quick Access Toolbar**, if it is not there then add it.

    ➢ Apply the **Critical** tasks filter.

    ➢ You will see only tasks that are on the Critical Path and their associated Summary tasks.

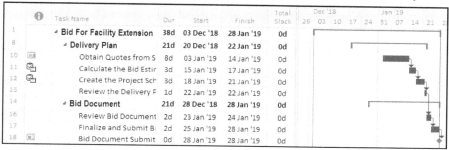

2. They would like to see all the tasks with float less than 10 days:

    ➢ Remove the previous filter.

    ➢ Use the **Filter** button, or the **View**, **Filter**, **More Filters...** to open the **More Filters** form.

    ➢ Create a new filter titled **Float Less Than 10 Days**,

    ➢ If you place a space at the start of the Filter name, the filter will be placed at the top of the list and make it easier to find filters you have created.

    ➢ Create a criteria to display a **Total Slack** of less than 10 days,

    ➢ Show the Summary tasks by checking **Show related summary rows** box,

    ➢ Show the filter in the menu, and apply the filter:

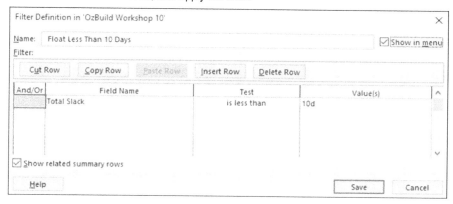

3. You should see that only the tasks with float less than 10 days are now displayed, the two tasks with 11 days Total Float are hidden.

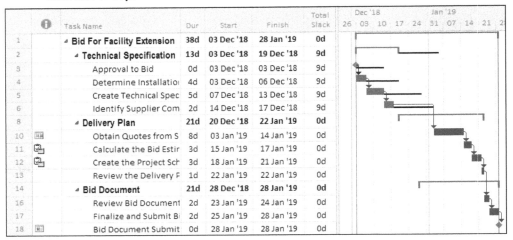

		Task Name	Dur	Start	Finish	Total Slack
1		▲ Bid For Facility Extension	38d	03 Dec '18	28 Jan '19	0d
2		▲ Technical Specification	13d	03 Dec '18	19 Dec '18	9d
3		Approval to Bid	0d	03 Dec '18	03 Dec '18	9d
4		Determine Installatio	4d	03 Dec '18	06 Dec '18	9d
5		Create Technical Spec	5d	07 Dec '18	13 Dec '18	9d
6		Identify Supplier Com	2d	14 Dec '18	17 Dec '18	9d
8		▲ Delivery Plan	21d	20 Dec '18	22 Jan '19	0d
10		Obtain Quotes from S	8d	03 Jan '19	14 Jan '19	0d
11		Calculate the Bid Estir	3d	15 Jan '19	17 Jan '19	0d
12		Create the Project Sch	3d	18 Jan '19	21 Jan '19	0d
13		Review the Delivery P	1d	22 Jan '19	22 Jan '19	0d
14		▲ Bid Document	21d	28 Dec '18	28 Jan '19	0d
16		Review Bid Document	2d	23 Jan '19	24 Jan '19	0d
17		Finalize and Submit B	2d	25 Jan '19	28 Jan '19	0d
18		Bid Document Submit	0d	28 Jan '19	28 Jan '19	0d

4. Management would like to see all the tasks with float less than, or equal to 10 days, **OR** contains the word "Bid."

> Copy the **Float Less Than 10 Days** filter,

> Assign a title to the filter: **Float Less Than 10 Days or Contains "Bid"**,

> Add the condition: **Or** Name (Task Name) contains **Bid**,

> Show in the menu, and

> Apply the filter.

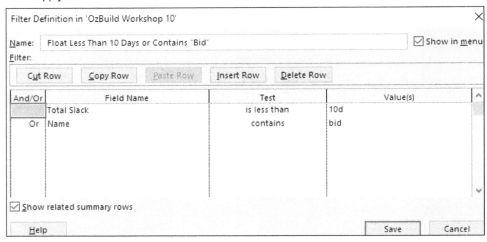

Filter Definition in 'OzBuild Workshop 10'

Name: Float Less Than 10 Days or Contains "Bid"   ☑ Show in menu
Filter:

| Cut Row | Copy Row | Paste Row | Insert Row | Delete Row |

And/Or	Field Name	Test	Value(s)	
	Total Slack	is less than	10d	
Or	Name	contains	bid	

☑ Show related summary rows

Help   Save   Cancel

---

5. You should find that one extra task is now shown.

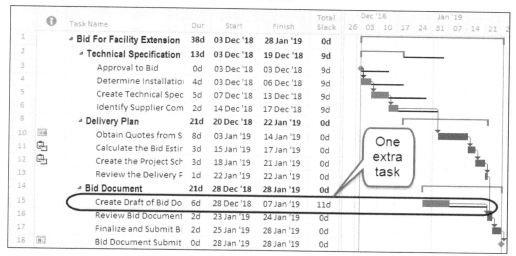

6. Copy the filter, title it **Float Less Than 10 Days and Contains "Bid"** and change the condition to **And** to see the difference between **Or** and **And** options,

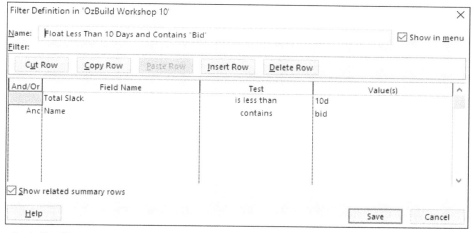

7. Apply the filter:

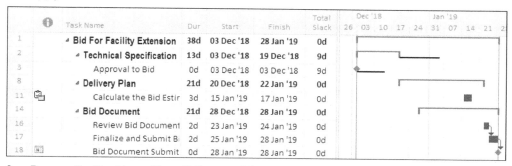

8. Remove the **Float Less Than 10 Days and Contains "Bid"** filter and display All Tasks by applying the **(No Filter)** or the **All Tasks** filter.

9. We now wish to create an **AutoFilter** that displays all the tasks containing the word "document".

10. Apply the (No Filter) option, which is the same as the All Tasks option used in other forms.

➢ Add the **Display Auto Filter** button to the **Quick Access Toolbar** if it is not displayed and use this to ensure the Auto Filters are activated and each column header has a down arrow as per this picture:

Task Name ▼

➢ Click on the down arrow in the Task Name header,

➢ Select **Filters, Custom...** to open the **Custom Auto Filter** form,

➢ Create a Custom Auto filter to contain tasks containing the word "document".

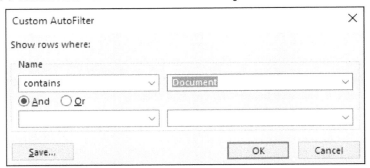

11. Now apply the **(No Filter)** or the **All Tasks** Filter to remove the **AutoFilter**.

12. Save your **OzBuild Bid** project.

---

13 TABLES AND GROUPING TASKS

Outlining was discussed earlier as a method of organizing detail tasks under summary tasks. There are alternative data fields and functions available in Microsoft Project for recording task information, then organizing, grouping and displaying task information:

- Text Columns, Custom Fields and Grouping
- Custom Outline Codes – covered in the **MORE ADVANCED SCHEDULING** chapter.
- User Defined WBS (Work Breakdown Structure) – covered in the **MORE ADVANCED SCHEDULING** chapter.

These functions are addressed in this book but are not examined in detail. These functions enable the presentation of the tasks under other project breakdown structures.

Topic	Menu Command
• Applying a Table to a View	• Click the **Select All** button (see paragraph 3.11) and right-click, or • Select **View**, **Data** group, **Tables** and select a Table from the list in the menu.
• Creating or Editing a Table	Select **View**, **Data** group, **Tables** and open the **More Tables** form.
• Create a Custom Field	Adding the ▥ **Custom Fields** button to your **Ribbon** or **Quick Access Toolbar**.
• Grouping	Select **View**, **Data** group, **Group by:**

13.1 Understanding Project Breakdown Structures

The main breakdown structure of a project is the WBS which is usually represented with Outlining. Some organizations have highly organized and disciplined structures with "rules" for creating and coding the elements of the structure.

Other Project Breakdown Structures are required to represent the breakdown of a project into other logical functional elements. Some clients also impose a WBS code on a contractor for reporting and/or claiming payments. The following list shows examples of such structures:

- WBS **Work Breakdown Structure**, breaking down the project into the elements of work required to complete a project.

- PBS **Product Breakdown Structure**, used in the PRINCE2™ Project Management Methodology.

- OBS **Organization Breakdown Structure**, showing the hierarchical management structure of a project.

- CBS **Contract Breakdown Structure**, showing the breakdown of contracts.

- SBS **System Breakdown Structure**, showing the elements of a complex system.

We will discuss the Text Columns, Custom Fields and Grouping functions available in Microsoft Project to represent these structures in your schedule.

13.2 Tables

A table selects and formats the columns of data to be displayed in a View. The formatting of tables is covered in the **FORMATTING THE DISPLAY** chapter.

- A table may be applied to one or more Views which display data in tables. This includes Views such as the Gantt Chart, Resource Sheet, Resource Usage, Task Sheet and Task Usage.

- There are two types of tables:
  - ➤ **Task** tables that are applied to **Task Views** and
  - ➤ **Resource** tables that are applied to **Resource Views**.

- When the View is active and you assign it a different Table, the View is permanently changed and the Table permanently associated with the View. Unlike Oracle Primavera software, the user does not have the option to save or not save changes to a view when another is selected.

 Formatting a table by adding or removing columns, etc., is editing the current table on a permanent basis. These changes will appear when the table is next applied and this will affect any View the table is associated with. It is therefore strongly recommended that each View be paired with a unique table of the same name. Consider carefully when adding or deleting columns from a table as the changes are permanent, unless you do not save your file.

13.2.1 Applying a Table to a View

A table may be applied to the active View by:

- Clicking the **Select All** button, the box above the row 1 number, then right-clicking the mouse to display a sub-menu with the table options, or

- Selecting **View**, **Data** group **Tables** and selecting a Table from the list in the menu.

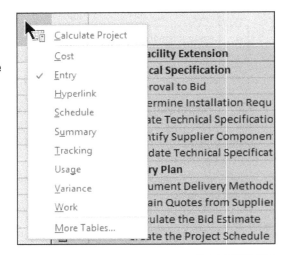

Both of these Table menus have a **More Tables...** option which will open the **More Tables...** form:

- Tables that have not been selected to appear on the menu list will be displayed here.

- Select a table from the list and click on Apply.

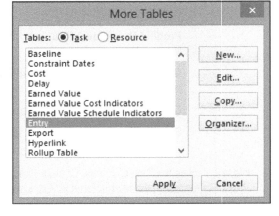

13.2.2 Creating and Editing a Table

A **Table** may be created or edited by opening the **View**, **Data** group **Tables**, **More Tables…**form:

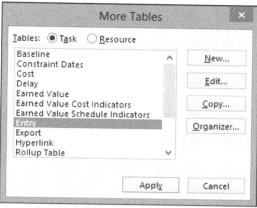

- **New…** creates a new table,

- **Edit…** edits an existing table, or

- **Copy…** creates a copy of an existing selected table.

- **Apply** – applies the table to a view.

All these buttons open the Table Definition form shown below:

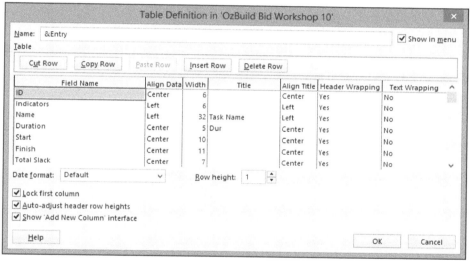

- The functions in this form are similar to those in many other forms. The functions that are unique to this form are listed below:

  - ➢ **Show in menu** – Decides if the Table is displayed in menus.

  - ➢ **Date format:** – Changes the format of the dates in this project and table only.

  - ➢ **Row height:** – Allows you to specify the row height for this table.

  - ➢ **Lock first column** – Ensures the first column is always displayed when scrolling to the right.

  - ➢ **Auto-adjust header row heights** – Automatically adjusts the header height when the width of the column is adjusted so the column text wraps.

  - ➢ **Show 'Add New Column' interface** shows the **New Column** column at the right of all columns. Clicking on this column creates a new column.

The **Date Format** selected in the **File**, **Options**, **General** tab is overridden by a date format selected in a Table. Therefore, if you have a project that requires a unique date format then the option of selecting a date format in a Table overrides the default on any computer for anyone who opens the project file.

13.3   Custom Fields

A **Custom Field** is an existing Microsoft Project field that may be:

- Renamed to suit your projects requirements,

- Tailored to display specific data in a specific format,

- Assigned a list, such as a list of values or people, that may be assigned from a drop-down list, or

- Assigned a formula for calculating data from other fields.

To create or edit a **Custom Field** the **Custom Fields** form must be opened by:

- Displaying the 📇 **Custom Fields** button Ribbon or Quick Access Toolbar, or

- Selecting 📇 **Custom Fields** when inserting a new column, or

- Selecting 📇 **Custom Fields** when right-clicking on an existing column.

Task fields may be used for:

- Recording additional information about Tasks (such as responsibility, location, floor, system)

- Recording additional information about Resources such as telephone number, address and skills.

- Formulas may be created to populate the fields with calculated data.

- Tasks or resources may be grouped using these fields.

These predefined fields fall into the following categories:

- Cost
- Date
- Duration
- Finish (date)
- Flag
- Number
- Start (date)
- Text
- Outline Code

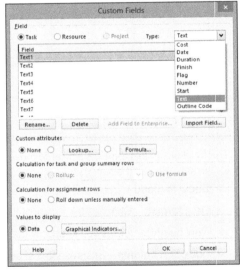

Both the title and content of these fields may be edited with options including:

- Rename... allows the renaming of the field name.
  - ➢ This new name is then available when inserting columns and is displayed in the column header.
  - ➢ After Custom Field is renamed, the new name will be displayed in the **Custom Fields** tab of the **Task** or **Resource Information** form and the appropriate information may then be entered in this form.

- Import Field... allows importing from other fields or project files.

- **Custom Attributes**:
  - ➢ **None** allows data to be entered into the field without any restrictions, this is similar to the way a User Defined Field works in Primavera P6.

  - ➢  opens the **Edit Lookup Table** where a table of values and descriptions may be entered. The Value is displayed in columns and Description in bands when the tasks are grouped by this field. These value may be copied and pasted from other applications such as Excel. This is similar to the way an Activity Code works in Primavera P6 and odes in Asta Powerproject.

  - ➢ [Formula...] allows the assigning of formulae for the calculation of field value from other task and project fields.

- **Calculation for task and group summary rows** specifies how Summary tasks calculate their value, such as Maximum, Minimum, Sum, None and Average. For example, the following options may be used:
  - ➢ A Start Date would select Minimum,
  - ➢ A Finish Date would select Maximum,
  - ➢ Cost would use Sum.

- **Calculation for assignment rows** determines if the field value is displayed only against the resource in Task Usage and Resource Usage fields or against the resource and assignment.

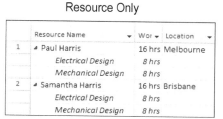

- **Value to display** allows the options of displaying the value in the cell or generating graphical indicators such as traffic lights. A very simple example is displayed below when the Number 1 Custom Field has been renamed Risk and three values entered and three different images displayed:

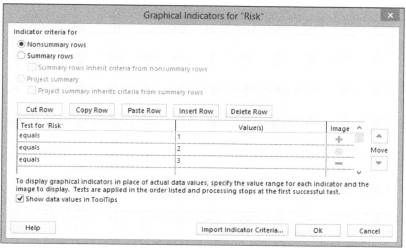

Outline Codes will be covered in more detail later in paragraph 24.7.1.

Oracle Primavera P3 and SureTrak software users will find the formatting options available when using Value and Description restrictive, because the description may not be displayed in columns and the value not displayed when Grouping. This is similar to the way Oracle Primavera P6 works.

13.4 Grouping

Grouping allows grouping of tasks under data items such as Customized fields, Durations, Constraints, etc. This function is particularly useful with schedules with a number of tasks and there is a requirement to work with a related group of tasks throughout a project. The picture below displays a simple project where the relationship between each Task is difficult to check by inspection of the Gantt Chart organized with Outlining by Phase:

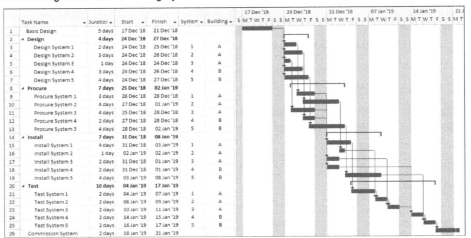

With the Grouping function it is possible to Group on a text field to reorganize the data. In this example, the schedule has been reorganized by the Text 1 - Building and Text 2 - System fields, which have been renamed using the **Custom Fields** form to System and Building. You may now clearly see the logic between the Items:

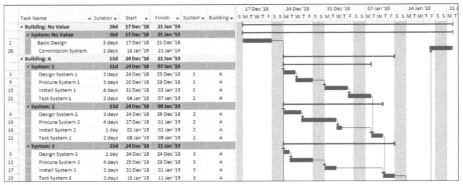

The first few characters of the field determine the sort order when tasks are grouped by a Text Field. To order items differently from the fields' text values, place a number or letter at the start of the description, or create a **Custom Outline Codes** which will take a little more effort but provide a more satisfactory result.

13.4.1 Group by: Function

The **View**, **Data** group, **Group by:** option allows you to Group scenarios in the same way as filters are created and saved:

- **Custom** are user defined groups,

- **Built-In** are system defined groups,

- **Clear Group** removes any grouping applied to a View,

- **New Group By...** opens the **Group Definition** form allowing the defining and saving of a new group scenario,

- **More Groups...** opens the **More Groups** form allowing access to Groups that may not be listed in the **View**, **Data** group, **Group by:** list.

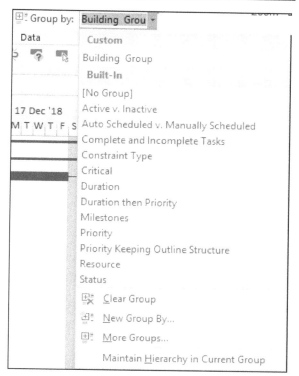

- **Maintain Hierarchy in Current Group** was a new feature to Microsoft Project 2010 and allows the user to see the Outline hierarchy when Grouping. The picture below displays the effect of grouping by two Custom Text columns, with the lower picture having this option turned on:

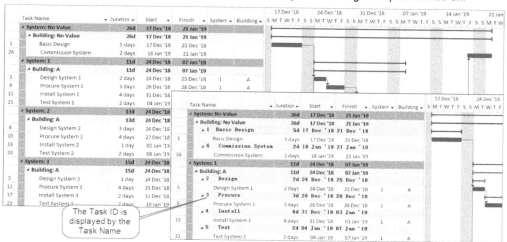

13.4.2 Using a Predefined Group

The Grouping function works in a similar way to Filters and Tables. A predefined Group may be assigned by:

- Selecting **View**, **Data** group, **Group by:**

- Then either:
  - ➢ Selecting a grouping from the list, or
  - ➢ Selecting **More Groups…** to open the **More Groups** form and then selecting one from the list after clicking on the **Task** or **Resource** radio button.

13.4.3 Creating a New Group

Create a new Group by:

- Selecting **View**, **Data** group, **Group by:**,

- Selecting **More Groups…** to open the **More Groups** form,

- Clicking on the [New…] button to open the **Group Definition** form,

- Now create a "Grouping" which may be reapplied at a later date or copy to another project using **Organizer**.

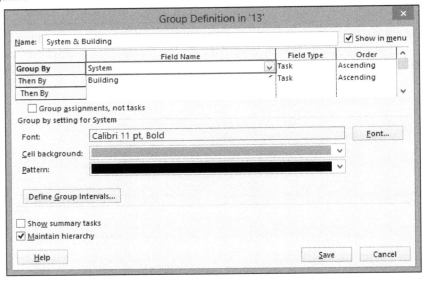

- The **Define Group Interval** form is available with many **Group By** options, such as Start or Finish, and allows further formatting options by defining the intervals of the banding.

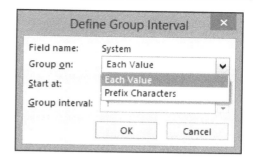

 Grouping is similar to the Oracle Primavera and Asta Powerproject's Group and Sort function. It is possible to mimic this Oracle Primavera function using the text columns as Task Code dictionaries. Projects converted from Oracle Primavera software format often translate Primavera Task Codes to Microsoft Project's Text fields. After conversion, the project may be **Grouped** by Text fields. **Custom Outline Codes** may produce a better result as bands may be ordered with this function.

13.4.4 Grouping Resources in the Resource Sheet

Resources may be created in the **Resource Sheet**. Then the resources may be grouped by a number of attributes. The standard options are shown below:

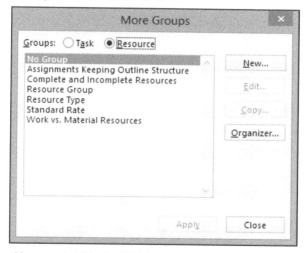

Resources are covered in more detail in the **RESOURCES** chapters.

There are many uses for Grouping Resources which may be used in conjunction with Customized Fields:

- A project hierarchical organizational structure may be created using these Customized Outline Codes and resources summarized under this hierarchical structure in the Resource Sheet.
- Resources details such as skill, trade, address, office, department and telephone number may be recorded in Customized Fields and the resources grouped by this data.

13.5　Workshop 11 - Reorganizing the Schedule

Background

We want to issue reports for comment by management. We will group the tasks by their float value and show the WBS columns. We will also look at the Outline Codes and then Group the Tasks by the people responsible for the work, which we will enter into a text column.

Assignment

1. Grouping – to group tasks without float
2. From the **View**, **Data** group:
   - ➢ Ensure there is no filter applied.
   - ➢ Apply the **Entry** table and ensure the **Total Slack** column is displayed.
   - ➢ Create a new Group titled **Total Float** and group the tasks by **Total Slack**.
   - ➢ Check the **Show in menu** option, **DO NOT** show Summary tasks and apply.

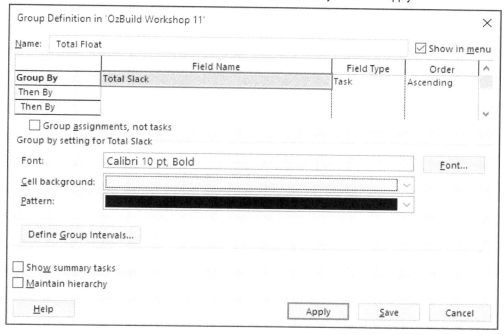

> ➤ All the tasks with zero days' float are grouped at the top under the heading **Total Slack: 0 days**.

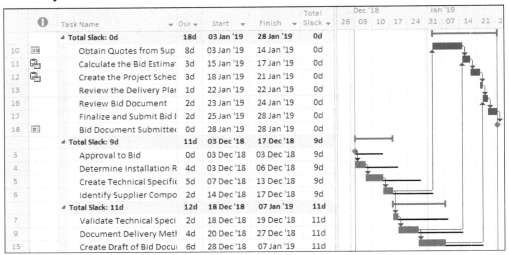

NOTE: You may format or remove the summary bars or add text to them by opening the **Bars** form and editing the **Group By Summary** bar text tab.

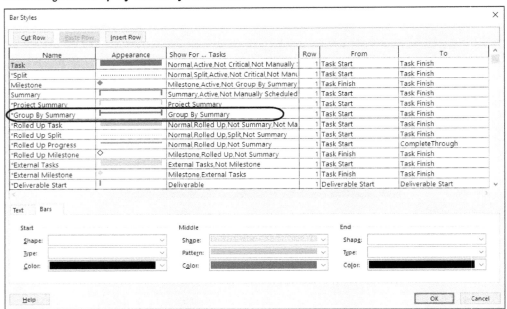

3. Grouping by Responsibility

➤ Remove the previous grouping by selecting [No Group],

➤ Use the **Add New Column** command to add a **Text 1** as a new column,

➤ Drag this new column so it is beside the Task Name

➤ Right Click on the column heading, select **Custom Fields** and use the **Rename** command to rename the column as **Responsibility**,

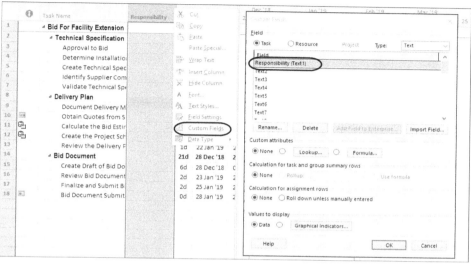

➤ Assign the Responsibilities in the table below, use Copy & Paste cells:

ID	Task Name	Responsibility
2	**Technical Specification**	
3	Approval to Bid	
4	Determine Installation Requirements	Scott Morrison - Engineering
5	Create Technical Specification	Scott Morrison - Engineering
6	Identify Supplier Components	Angela Lowe - Purchasing
7	Validate Technical Specification	Scott Morrison - Engineering
8	**Delivery Plan**	
9	Document Delivery Methodology	Scott Morrison - Engineering
10	Obtain Quotes from Suppliers	Angela Lowe - Purchasing
11	Calculate the Bid Estimate	Carol Peterson - Bid Manager
12	Create the Project Schedule	Carol Peterson - Bid Manager
13	Review the Delivery Plan	Carol Peterson - Bid Manager
14	**Bid Document**	
15	Create Draft of Bid Document	Carol Peterson - Bid Manager
16	Review Bid Document	Carol Peterson - Bid Manager
17	Finalize and Submit Bid Document	Carol Peterson - Bid Manager
18	Bid Document Submitted	

➤ Create a Group titled **Responsibility** by clicking the **Group by: button** 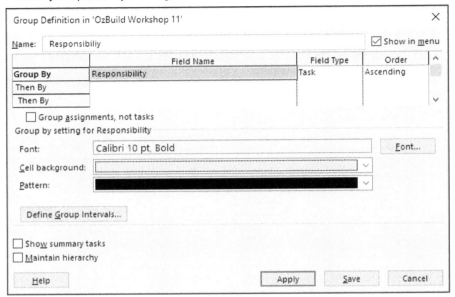 and grouping the tasks by Responsibility showing in the menu but do not show Summary tasks.

➤ Select **Apply**,

➤ See the answer below:

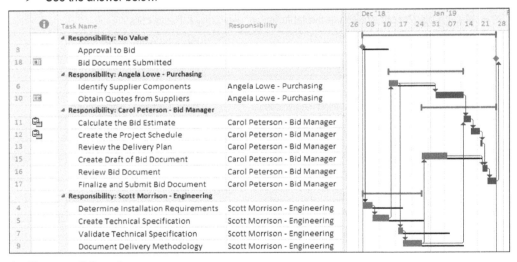

4. Remove all Grouping.

5. Hide the **Responsibility** column.

6. Save your **OzBuild Bid** project.

14 VIEWS AND DETAILS

A **View** is a function where the formatting such as **Grouping**, **Table**, **Filter**, **Print Settings** and **Bar** formatting are saved as a **View** and reapplied later.

All Views, except the **Timescale** view which is normally displayed above the Gantt Chart, may be applied by:

- Clicking the appropriate button from the **Ribbon View**, **Task Views** and **Resource Views** groups.

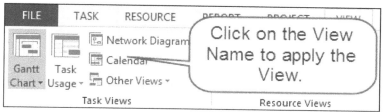

- Clicking on the down arrow ▼ by a View Name on **View**, **Task Views** and **Resource Views** groups to open a submenu and then click on the View name to apply the View:

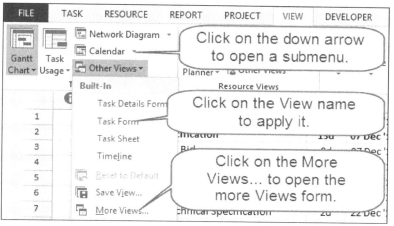

- There are more views available than the Views listed on the **Ribbon** menu. These may be applied from the **More Views...** form. This may be accessed from all the **View**, **Task Views** and **Resource Views** groups, by placing a button on the **Quick Access Toolbar** and, possible the simplest way, to right click in the dark band on the left-hand side of the active pane:

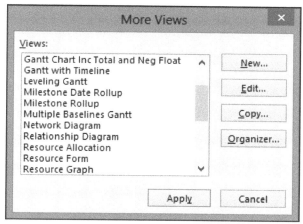

- A View may be displayed by right-clicking in the bar on the left had side of the screen which will open a menu with a list of Views:

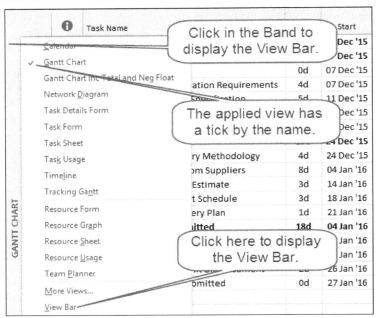

- A View may also be applied by displaying the **View Bar** and clicking on a **View** button. The **View Bar** is displayed as shown in the picture below.

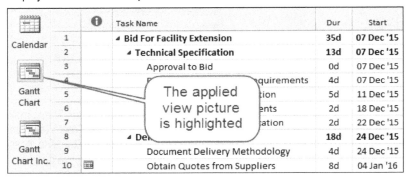

- There is also a button on the **Task**, **View** group that displays a list of available task related Views:

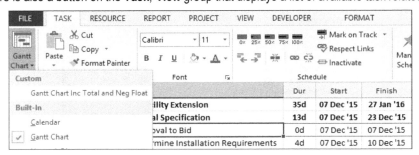

The **Timeline** view is shown above the Gantt Chart by selecting **View**, **Split View** group and checking the **Timeline** check box. This View is covered in more detail at the end of this chapter.

The views displayed in the **Ribbon** menu and displayed on the **View Bar** with a standard install of Microsoft Project are listed below:

View Name	Ribbon Command	Comments
Calendar	**Task**, **View** group, **Calendar**	This may be applied to the top window only and displays the tasks overlaid on a calendar.
Gantt Chart	**Task**, **View** group, **Gantt Chart**	Displays a Table on the left of the screen and Bar Chart on the right.
Network Diagram	**Task**, **View** group, **Network Diagram**	This view is a graphical PERT-style display and is covered in the **NETWORK DIAGRAM VIEW** chapter.
Relationship Diagram	Not on the **Ribbon** as a default.	This is a graphical view showing the relationship among tasks. Usually placed in the bottom pane.
Resource Form*	**Resource**, **View** group, **Resource Form**	This displays one Resource at a time and list the tasks assigned to the resource. Usually placed in the bottom pane.
Resource Graph*	**Resource**, **View** group, **Resource Graph**	A split table displaying the Resource name on the left-hand side and resource usage in bar chart format on the right-hand side.
Resource Sheet*	**Resource**, **View** group, **Resource Sheet**	A single table used to create resources and displaying Resource information such as costs and calendars.
Resource Usage*	**Resource**, **View** group, **Resource Usage**	Similar to the Task usage, displays a Table with Resources with associated Tasks on the left and the Resources usage data apportioned over time on the right.
Task Form	**Task**, **View** group, **Task Form**	This is the form that is normally displayed in the bottom pane when the Gantt Chart view is split.
Task Sheet	**Task**, **View** group, **Task Sheet**	This is the same as a Gantt Chart view, button bars are displayed.
Team Planner	**Task**, **View** group, **Team Planner**	This was new to Microsoft Project 2010 and available only with the Professional Version. The view allows a graphical view of Resource assignment to Tasks.
Timeline	**Task**, **View** group, **Timeline**	This was a new view to Microsoft Project 2010 which displays the whole project graphically above the Gantt Chart.
Tracking Gantt	**Task**, **View** group, **Tracking Gantt**	This is a Gantt Chart view for entering progress which uses the tracking table and displays the Baseline.

* The Views containing resource information are covered in more detail in the **UPDATING PROJECTS WITH RESOURCES** chapter.

The **Calendar** view is not covered in detail in this book. In the Calendar view and when there are many tasks on one day, all the tasks may not be displayed depending on the scale of your screen and or paper size when you are printing the Calendar View.

The **Network** view is covered in the **NETWORK DIAGRAM VIEW** chapter.

14.1 Understanding Views

There are two types of Views:

- A **Single View** is normally applied to the active pane only; when the pane is split, or the whole screen when not split.

- A **Combination View** is comprised of two **Single Views**, one displayed in the top pane and one in the bottom pane.

The top pane is usually used to display all data and the bottom pane display specific data relating to the selected task or tasks in the top pane.

All **Single Views** may be applied to the top pane and when the window is split, most may be applied to the bottom pane. When a **Single View** is applied to the bottom pane, it will only display the information attributed to the task that is highlighted in the top pane.

When a **Single View** is applied to the top or bottom pane, the other pane is left with the contents of the previous **Combination View**.

A **View** is based on a **Screen** when it is created. The **Screen** may not be changed after the **View** is created. There are 16 **Screens**:

• Calendar	• Gantt Chart
• Network Diagram	• Relationship Diagram
• Resource Form	• Resource Graph
• Resource Name Form	• Resource Sheet
• Resource Usage	• Task Details Form
• Task Form	• Task Name Form
• Task Sheet	• Task Usage
• Team Planner – New to 2010 Professional Version Only	• Timeline – New to 2010

The Table in **APPENDIX 1 – SCREENS USED TO CREATE VIEWS** lists the Screens and provides further detail about formatting the screens. A few important points to consider are:

- The **Calendar** screen may only be displayed in the top pane.

- The **Gantt Chart** and **Network Diagram** are often best displayed in the top pane and have limited use when displayed in the bottom pane.

- **Relationship Diagram** is best displayed in the bottom pane.

- The **Forms** are best displayed in the bottom pane and convey **Task** or **Resource** information about a task that is highlighted in the top pane. **Forms** may be further formatted by right-clicking in the form to open a menu.

- The **Task Sheet** is similar to the **Gantt Chart** but does not display bars and is best displayed in the top pane.

- The **Resource**, **Task Usage** and **Team Planner** screens display resource information and are further discussed in the **RESOURCE OPTIMIZATION** chapter.

- **Resource Graph** and **Resource Tables** are best displayed in the bottom pane.

14.2   Creating a New View

A new View may be created by copying and editing an existing View, or creating a new View.

14.2.1   Creating a New Single View

To create a new Single View:

- Open the **More Views** form, which is available from many view buttons,

- Click on the 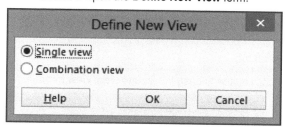 button to open the **Define New View** form:

- Click on the **Single view** radio button to open the **View Definition** form:

- You may now enter:

    ➢ The new View name in the **Name:** box by overtyping the "**View 1**" name assigned by default by Microsoft Project,

    ➢ Select the **Screen:** View. (see **APPENDIX 1** for more information on the screens). A list of available screens is displayed below:

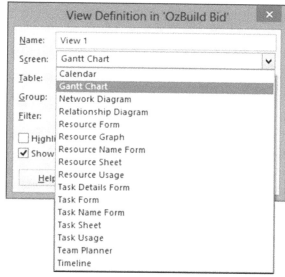

    ➢ Select the **Table:** you want to apply with the view.

    ➢ You may not want to display your tasks using the Outline display mode. The **Group:** function allows you to group your tasks by other data items such as a **Text Column** or **Constraint Type** and is covered in the **TABLES AND GROUPING TASKS** chapter.

    ➢ You may also select a filter to apply with the View from the **Filter:** drop-down box.

    ➢ Click the **Highlight filter** box to highlight the tasks that meet the filter specification as opposed to applying the filter.

    ➢ Click on the **Show in menu** box to display the View in a menu.

- Click on the [OK] button to create the new **View**.
- From the **More Views** form click on the [Apply] button to apply the new **View**.

 Inserting a space before the View name will put the View to the top of the list making it simpler to identify those Views that you have created or edited.

14.2.2 Creating a Combination View

Before creating a new Combination View the two Views that are to be displayed in the top and bottom pane must have been created:

- Open the **More Views** form,
- Click on the [New...] button to open the **Define New View** form:

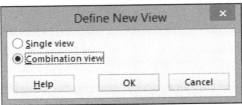

- Select **Combination view** and click [OK] this will open the **View Definition** form:

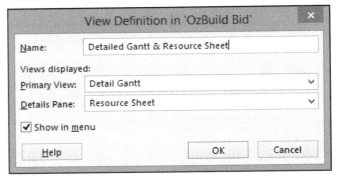

- Enter your View name and select the Views you want to display in the top and bottom pane,
- Click on **Show in menu** to display the View in the menu and on the **View Bar**, and
- Click on the [OK] button to create the new **View**.

 A useful technique is to create a View, and associated Table and Filter, with the same name and keep them together as a set. For example, a View created to show critical tasks could have a View, Table and Filter all titled "Critical."

14.2.3 Copying and Editing a View

To copy an existing View:

- Open the **More Views** form:

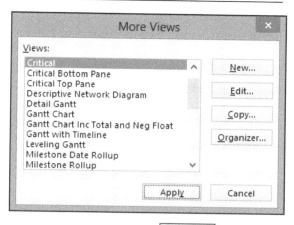

- Select the View you want to copy from the drop-down list and click on the [_Copy..._] button to open the **View Definition** form. This form is different for Single and Combination Views. The example below is a **Single View** created from the **Gantt Chart** Screen:

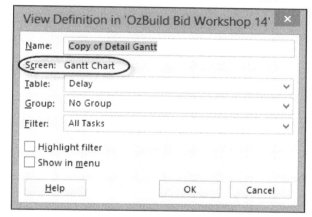

- The **Screen:** may not be changed in an existing or copied view and the screen option is shown as gray. See picture above.

- Change the name and any other parameters and then click on [OK] to create the **View**.

14.2.4 Copying a View to and from Another Project

Views may be copied from one project to another with the **Organizer** function. The **Organizer** function may be accessed by selecting:

- [Organizer...] from the **More Views** form, or

- Placing the **Organizer** button on the **Ribbon** or the **Quick Access Toolbar**.

 If a View is copied to another project then the associated Tables, Filters and Grouping may also need to be copied.

14.3 Details Form

Details forms are the third level of formatting that may be assigned in some views. An extensive list of the Details forms is outlined in **APPENDIX 1 – SCREENS USED TO CREATE VIEWS**.

Each view has a number of Details options, which tends to make this aspect of Microsoft Project difficult for all levels of users.

The **Details** forms may be selected in the bottom pane by right-clicking in the active pane to open a menu. Examples of Details forms are displayed below:

- **Task Form** with Resource Costs Details form:

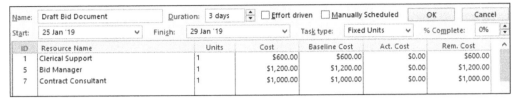

- **Task Details Form** with Predecessors and Successors Details form which displays more information than the **Task Form**:

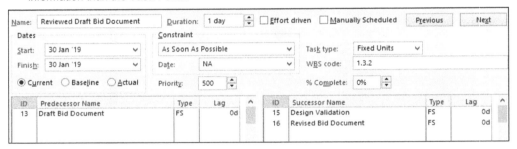

14.4 Timeline View

14.4.1 Microsoft Project 2013 Timeline

This view, new to Microsoft Project 2010, allows you to clearly understand how much of the Gantt Chart is displayed on the screen. The picture below demonstrates that about one-third of the project is displayed in the Gantt Chart.

The **Timescale** view is shown above the Gantt Chart by selecting **View**, **Split View** group and checking the **Timeline** check box. Right-clicking in the **Timeline** will show a menu, as displayed in the picture below:

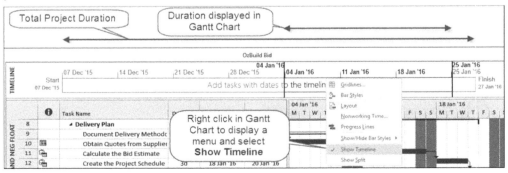

This publication is only sold as a bound book, no parts may be reproduced by any means, e.g. electronic, video or print.

© *Eastwood Harris Pty Ltd*　　　　172

A task may be added to the **Timeline** view by selecting the **Add to Timeline** found in many menus. Important tasks that need to be highlighted should be considered for this type of display. In the picture below the Start and Finish Milestones and Obtain Quotes have been added to the Timeline view:

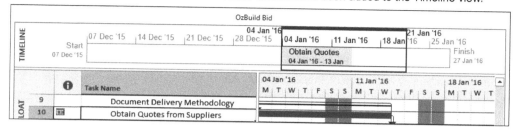

14.4.2 Understanding Microsoft Project 2016 Timeline Bars

Microsoft Project 2016 introduced the ability to:

- Add multiple Timelines,

- Change the start and Finish date of the Timeline.

- Format the Colors of each individual Timeline and

To add a second Timeline:

- Left click on the Timeline bar, then select **Format**, **Insert**, **Timeline Bar**:

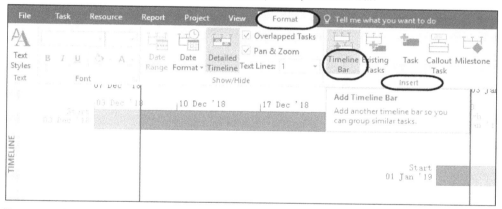

Change the Start and Finish date of the Timeline.

- Right Click on the Timeline Bar to open a menu and

- Select **Date Range**

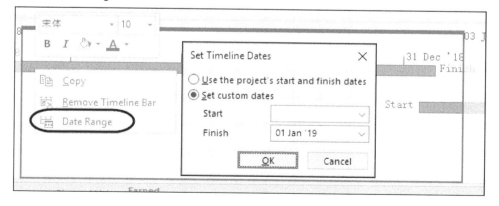

To Remove a second Timeline Bar:

- Select the Timeline Bar,

- Right Click on the Timeline Bar to open a menu and

- Select **Remove Timeline Bar**

14.4.3 Formatting Microsoft Project 2016 Timeline Bars

Before a Timeline may have further formatting applied, it must be made a Detailed Timeline Bar.

To create a Detailed Timeline Bar:

- Select the Timeline Bar,

- Move the mouse so it is no longer on the bar,

- Right Click to display the menu that is shown in the picture below,

- Select **Detailed Timeline:**

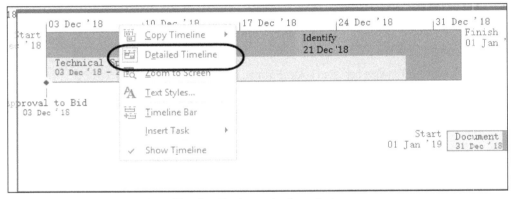

Once a task has been added to a Timeline Bar it may be formatted:

- To Change the color of a Timeline Bar right click and use the formatting menu to change the font and color,

- A task may be made a **Callout** and dragged around the View by right clicking and selecting **Display as a Callout**:

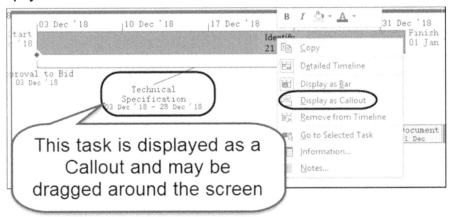

14.5 A Logical Process for Developing a View

A View may be used to create a management report or to manipulate a schedule. The following tips should be considered when creating a view that will be used for reporting:

- It is important that you plan the process you wish to use to create Views. A small amount of planning at the start of a project will save a large amount of time later in a project lifecycle and avoid changing a number of views.

- When there is a requirement to create a number of Views that will all have the same headers and footers when printed out, then it is best to create one View and spend time ensuring that the print settings are correct. Then copy the View, which copies the print settings, to create the remainder of the Views.

- It is also advisable to name the Filters, Table and Groupings with the same name so that it can easily be understood where Filters, Table and Groupings belong.

- When there are a large number of views to be created a numbering or coding system should be considered and used as a prefix to their names which will group and order the Filters, Table and Groupings.

- Filters, Table and Groupings that are part of Views should not be shown in the menu so their names not inadvertently changed.

The following process is suggested for creating a new View;

- Collect the View requirements,

- Plan the naming/coding system to be used if there are a number of views to be created,

- Create any Custom fields required,

- Create the Table, Filter and Grouping to be used with the View,

- Create the View and assign the Table, Filter and Grouping,

- Create the print headers and page setup,

- Format the Timescale, Gridlines, Bars,

- Apply any special sorting or font formatting,

- Save the file.

14.6 Workshop 12- Organizing Your Data Using Views and Tables

Background

Having completed the schedule you want to report the project schedule with different views.

Assignment

Display your project in the following formats, noting the different ways you may represent the same data. Open your **OzBuild Bid** project from the previous workshop to complete the following exercise.

1. Right click in the bar on the left hand side of the task numbers, display the **Calendar** view and scroll to December 2018. Right-click in different parts of the screen and view the menu options. When there are many tasks on one day, they may not all be displayed, depending on the scale of your screen and/or paper size when you are printing the Calendar View.

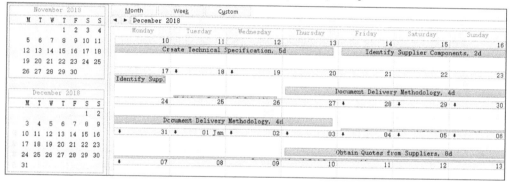

2. Display the **Network Diagram** view and zoom to 50%, using **View**, **Zoom** group, **Zoom** or the ⊕ ⊖ buttons.

3. Scroll around the schedule and then summarize the task by clicking on the ⊞ and the ⊟.

4. Display the **Gantt Chart** view and split the screen.

5. Display the **Relationship Diagram** view in the bottom pane by first making the bottom pane active and then right clicking in the band on the left of the screen to open the **More Views** form. Scroll around the schedule by clicking on tasks in the top pane and review the changes in the bottom pane.

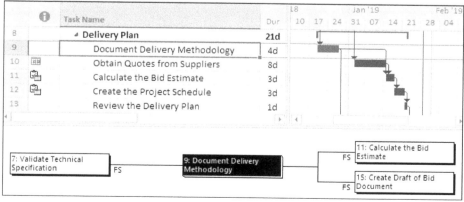

6. Display the **Task Details** view in the bottom pane. If this menu is not immediately available Select, **More Views**. Notice the additional fields that are available in this form compared to the **Task Information** form.

7. Apply each one of the following **Details** forms to the **Task Details** view in the bottom screen by right-clicking in the bottom pane and observe the different information for each option.

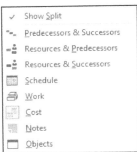

8. Display the Timeline View by selecting the **View**, **Split View** group and check the **Timeline** box.

9. Change the timescale to expand the Gantt Chart so not all of project timescale may be viewed. Observe the Timeline duration displayed in the Gantt Chart and the highlighted section of the Timescale are the same:

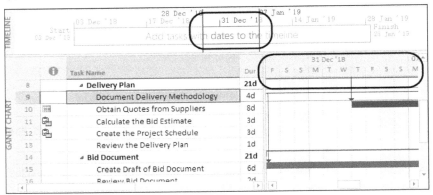

10. Change the timescale back to the previous settings, so you may see the whole Gantt Chart.

11. Add both Milestones and the **Obtain Quotes from Suppliers** task to the **Timeline** by right clicking and selecting **Add to Timeline**.

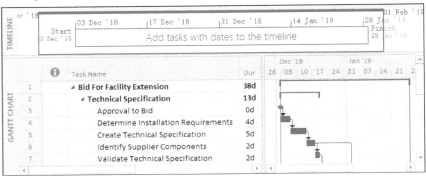

12. Remove the **Timeline** by right clicking in the Gantt Chart and click on **Show Timeline**.

13. Display the **Gantt Chart** view, with the **Entry** table and split the screen.

14. Ensure **Critical** tasks are displayed by checking the **Format**, **Bar Styles**, **Critical** box.

15. Apply the following Tables to the **Gantt Chart** view using **View**, **Data**, **Tables**:

   ➢ **Schedule**, then

   ➢ **Tracking**, then

   ➢ **Entry**

16. We now will create a Combination View titled **Critical** to show only Critical Tasks in the top pane with the Task Details form in the bottom pane. We will create the following:

   ➢ A Critical Table

   ➢ A Critical View for the Top Pane

   ➢ A Critical View for the Bottom Pane and

   ➢ A Critical Combination View, with the Top Pane Critical View and Bottom Pane Critical View.

17. To create the Critical Table:

   ➢ Select **View**, **Data** group, **Tables, More Tables** and copy the **Entry** table and create a table named **Critical** with the following columns:

   ➢ ID, Indicators, Name, Duration, Start, Finish, Total Slack and Free Slack.

   ➢ Do not check **Show in the menu**,

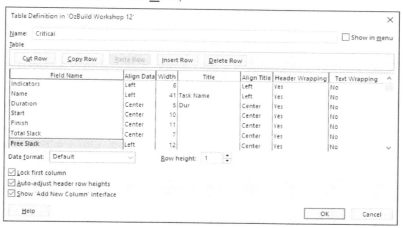

18. To create the Critical View for the Top Pane:

   ➢ Copy the **Gantt Chart** view by selecting **View, Task Views** group and from any menu select **More Views**,

   ➢ Rename the view **Critical Top Pane**,

   ➢ Select the **Critical** Filter,

   ➢ Select the **Critical** Table,

   ➢ Ensure the **Show in menu** is not selected,

   ➢ Close the forms without applying the view.

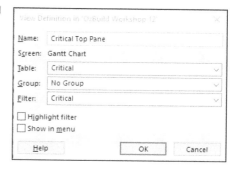

---

19. To create the Critical View for the Bottom Pane:

> Select **View, Task Views** group and from any menu select **More Views**,

> Create a new **Single** view titled **Critical Bottom Pane** using the **Task Details** form ,

> Select **All Tasks** filter,

> Do not **Show in menu**, as per the picture on the right:

> Close the form without applying the view.

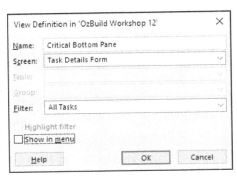

20. To create the Combination View:

> Select **View, Task Views** group and from any menu select **More Views**,

> Create a new **Combination** view titled **Critical Combination** as per the picture on the right,

> Select **Show in menu**,

> Close the forms without applying the view

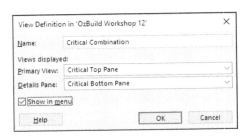

21. Apply the **Critical Combination** I view and you should see that the Critical Filter and Critical Table are applied in the Top Pane and Task Details Form in the bottom pane:

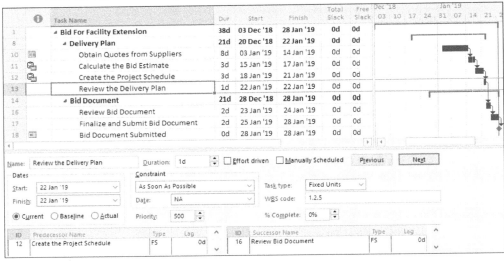

22. Apply the **Gantt Chart** view and remove the split

23. Save your **OzBuild Bid** project.

15 PRINTING AND REPORTS

You are now at the stage to print the schedule so the project team may review and comment on it. This chapter will examine some of the many options for printing your project schedule.

There are two tools available to output your schedule to a printer:

- The **Printing** function prints the data displayed in the current Active View.

- The **Reporting** function prints reports, which are independent of the current View. Microsoft Project supplies a number of predefined reports that may be tailored to suit your own requirements.

It is recommended that you consider using a product such as Adobe Acrobat to output your schedule in pdf format. There are many low cost pdf creating software packages available on the internet. You will then be able to email high quality outputs that recipients may print or review on screen without needing a copy of Microsoft Project.

It is also easier to fit a Gantt Chart View which has many columns on an A3 or 11"x17" than an A4 or Letter sized output. Therefore it is often best to print to an A3 or 11"x17" pdf and then print to an A4 or Letter sized piece of paper using the pdf printing software.

15.1 Printing

Only the active View may be printed when a screen is split. The writing in the active view text is highlighted on the left-hand side of the screen. Views created from **Forms** (for example, the Task Form) may not be printed, so the printing options will be shown in grey when the forms are active.

Print settings are applied to the individual Views and the settings are saved with the currently displayed View.

In Microsoft Project 2013 the printing commands have been consolidated under the **File**, **Print** command.

Microsoft Project sometimes makes it difficult to print a Gantt Chart on one page. The author recommends adjusting the timescale so the whole project fits into half the screen before selecting **Print Preview**, which makes this process simpler.

Each time you report to the client or management, it is recommended that you save a complete copy of your project and change the name slightly (perhaps by appending a date to the file name or using a revision or version number) or create a subdirectory for this version of the project. This allows you to reproduce these reports at any time in the future and an electronic copy is available for dispute resolution purposes.

15.2   File, Print Form

The Microsoft Project 2013 **File**, **Print** form has replaced the earlier versions **Print Preview** and **Print** form. To preview the printout:

- Select **File**, **Print**, or

- Click on the **Quick Access Toolbar** 🔍 **Print Preview** button or the **Ribbon**.

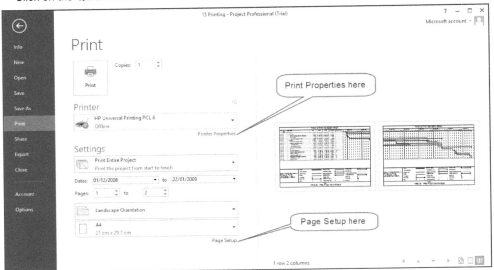

15.2.1   Print and Printer Heading

This heading allows the selection of:

- The printer and number of copies

- Printer properties.

15.2.2   Setting Heading

This heading has several options that allow the selection of the dates, pages to be printed and which columns to be printed:

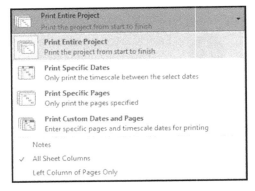

- **Print Entire Project** – prints the whole Gantt Chart

- **Print Specific Dates** – prints the selected dates. Changing the dates will change the option.

- **Print Specific Pages** – allows the selection of specific pages to be printed when multiple pages are displayed.

- **All Sheet Columns** – when unchecked will only print the visible columns displayed in the Gantt Chart.

- **Left Column of Pages only** – will print only the left pages of a printout when they are more than one page wide.

15.2.3 Buttons at the Bottom Right Hand Side of Print Form

The following paragraphs describe the functions of the buttons at the bottom right-hand side of the Print Preview screen from left to right:

- The first four buttons on the left, [◀ ▲ ▼ ▶], allow scrolling when a printout has more than one page.

- The next three buttons, [⊞], [▤] and [▤], display one or all pages and the actual size, respectively.

- Clicking on the print preview to zoom in and out, the cursor will change into a 🔍 or a 🔍.

15.3 Page Setup

Page Setup is unique to each View. To open the **Page Setup** form:

- Click the **Page Setup** link at the bottom of the **Settings** section, or

- Place a [▣] **Page Setup** button on the **Quick Access Toolbar** or **Ribbon**:

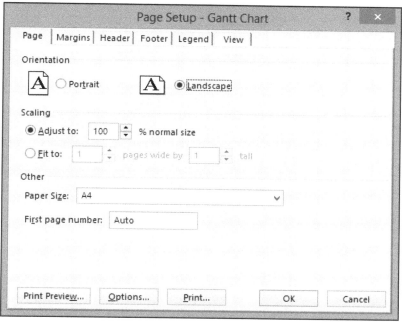

The **Page Setup** form contains the following tabs:

- Page
- Margins
- Header
- Footer
- Legend
- View.

Depending on the View being printed, some options in the **Print Setup** form will be unavailable and shown in gray.

The [Options...] button opens the printer properties form.

When the **Page Setup** form is opened from the **Quick Access Toolbar** or **Ribbon** then the
Print Preview... button becomes available and takes you to the **Print** form.

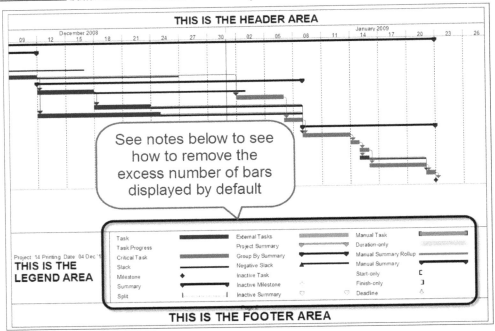

The picture above was created using the Microsoft Project Global.mpt and by default it has a large number of bar styles displayed in the legend. It is better not to delete these bars because if a task is created but the bar type has been deleted you will not see the bar. It is better to place an asterisk "*"in front of bar **Name** in the **Bar Styles** form so the bar is not displayed in the **Legend**, see the picture below:

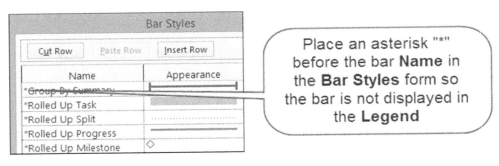

15.3.1 Page Tab

The Microsoft Project options in the **Page** tab are:

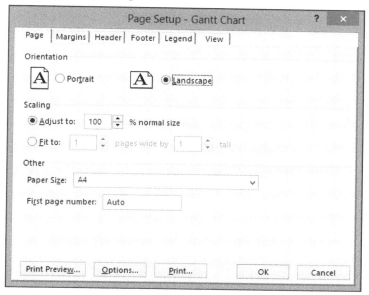

- **Scaling** allows you to adjust the number of pages the printout will fit onto:

  ➢ **Adjust to:** – Allows you to choose the scale of the printout that both the text is scaled to. Microsoft Project will calculate the number of pages across and down for the printout.

  ➢ **Fit to:** – Allows you to choose the number of pages across and down and Microsoft Project will scale the printout to fit.

- The **First page number:** – Allows you to choose the first page number of the printout. This is useful when enclosing the printout as an attachment or an appendage to another document.

- Pages are numbered down first and then across:

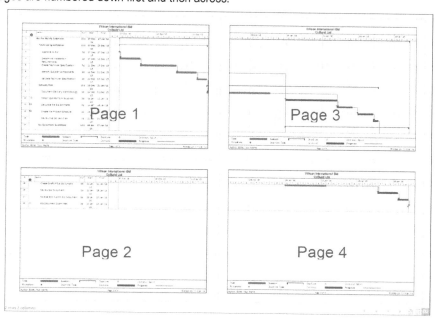

15.3.2 Margins Tab

With this option you may choose the margins around the edge of the printout.

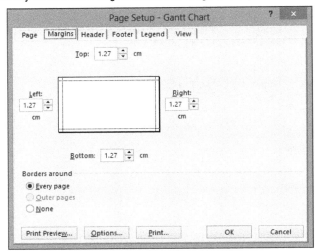

- Type in the margin size around the page. It is best to allow a wider margin for an edge that is to be bound or hole-punched; 1 inch or 2.5 cm is usually sufficient. The units of measurement, in. or cm, are adopted from the operating system **Control Panel**, **Regional and Language Option** settings.

- Microsoft Project will print a border around the Gantt Chart bars and Columns. **Borders around** places a border line around the outside of all text in the Headers and Footers. There are three options:

  ➤ **Every page** – This places a border around every page.

  ➤ **Outer pages** – This capability is only available with a **Network Diagram** view. It allows you to join all the pages into one large printout with a border only on the outside edge of all the pages once they are joined up.

  ➤ **None** – Does not place a border on any sheets of the printout.

15.3.3 Header, Legend and Footer Tabs

Headers appear at the top of the screen above all schedule information and footers are located at the bottom, below the **Legend** if it has been selected to be displayed. Both the headers and footers are formatted in the same way. We will discuss the setting up of footers in this chapter.

Click on the **Footer** tab from the **Page Setup** form. This will display the settings of the default footers and headers.

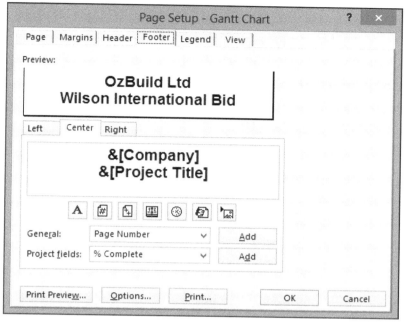

- **Preview:** – The box at the top of the form shows how your Footer will be displayed.
- Text may be placed in all three positions in the Footer and Header – on the Left, Center. Right-click on the required tab to insert new text or remove existing text in the box below.
- Text may be added by using a number of methods:
  - ➢ Freeform text may be typed into the footer box below the **Alignment:** box,
  - ➢ Add data in the footer by clicking one of these buttons, ⊞ **Page Number**, ⊡ **Total Page Count**, ⊞ **Current Date**, ⊗ **Current Time**, ⊟ **File Name** or ⊡ **Insert Picture** (for example a corporate logo),
  - ➢ To insert Microsoft Project field-type information to your footer, click on [Add] to insert the field into the footer. You may select from the drop-down box to the right of **General:** or **Project fields:**,
  - ➢ To format the text you must first highlight the text and then click on the **Format Text Font** button [A] to open the **Font** form.
  - ➢ [Print Preview...] — Returns you to the **Print Preview** form.
  - ➢ [Options...] — Opens the **Printer Properties** form.
  - ➢ [Print...] — Opens the **Print** form.
  - ➢ [OK] — Accepts the changes.
  - ➢ [Cancel] — Cancels any recent changes that have not yet been saved.

15.3.4 Legend Tab

The **Legend Text** is printed to the left of the **Legend**:

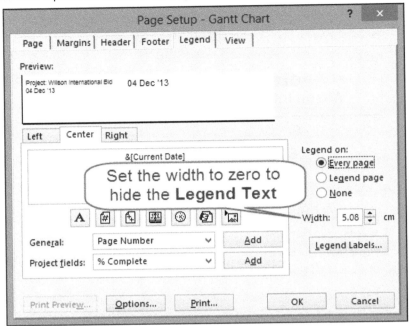

The **Legend Text** is formatted in the same way as the **Header** and **Footer**.

The **Legend** has three additional options not available in the **Header** and **Footer**.

- Click on the radio button below **Legend on:**
  - ➤ **Every page** will print the legend at the bottom of every page.
  - ➤ **Legend page** will print the legend on a separate page with no detailed schedule data, bars, or columns.
  - ➤ **None** will not print a legend.
- **Width:** sets the width of the **Legend Text**. Set the width to zero to hide the legend.
- Legend Labels... Opens the **Font** form for formatting the font of the text in the legend next to each of the bars and milestones.
- To hide a bar type in the Legend, type an "*" in front of the description in the **Bar Styles** form. This bar will still be displayed but will not be displayed in the Legend:

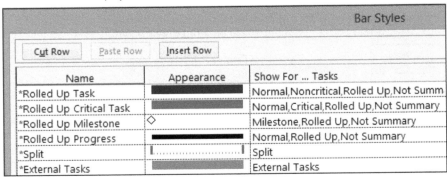

15.3.5   **View Tab**

The View tab has five options:

- **Print all sheet columns:** – This option applies to the left-most horizontal pages only when there is more than one page across a printout.

  ➤ When checked, this option will print all the sheet columns displayed by the current table in the Gantt Chart, even if the columns are hidden by the vertical divider in the normal view.

  ➤ When unchecked, this option will only print the completely visible columns.

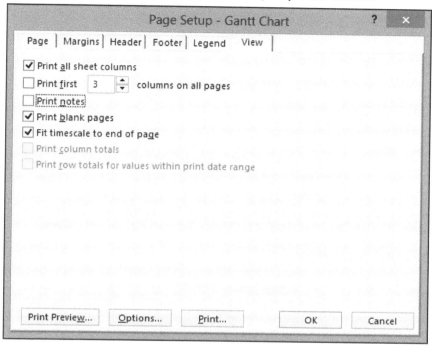

- **Print first** ... **columns on all pages** – Applies to all pages. The greater of the number of columns between this option and the options above will be printed on the first page.

  ➤ **When checked**, this will allow repetitiously printing the selected number of columns on all pages.

  ➤ **When unchecked**, this will not print any columns on the second and subsequent pages.

- **Print notes** will allow the printing of any **Notes** such as **Task Notes** on a separate page.

- **Print blank pages** will prevent the printing of all pages in the Gantt Chart, when there are no bars.

- **Fit timescale to end of page** will extend the timescale and associated bars so they fit to the end of a page, but sometimes you may find better results by turning this option off:

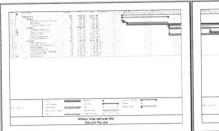 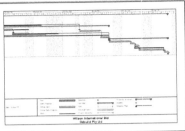

  ➢ Option is checked, bars and timescale extend to the end of a page:

 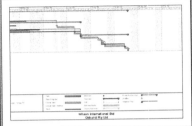

  ➢ Option is unchecked, the bars stop in the middle of a page:

- **Print row totals for values within date range** and **Print column totals** options become available when tabular data is selected for printing such as a Resource Sheet view.

15.4   Manual Page Breaks

Manual page breaks may be inserted but these did not print on the author's earlier installs of Microsoft Project. Also the **Print** form **Manual page breaks** check box from the earlier version of Microsoft Project was missing and could not be found on any list of commands. This issue has been resolved with software upgrades and page breaks now work with later versions of Microsoft Project 2013, but the **Manual page breaks** check box has not appeared, so manual page breaks may not be disabled and have to be deleted if not required.

Manual page breaks are inserted by:

- Placing the ⊙ **Manual Page Break** button on the **Ribbon** or **Quick Access Toolbar**,

- Highlighting the row above where a page break is required. Then click on the ⊙ **Manual Page Break** button. A dotted line will indicate the location of the manual page break.

- To remove a manual page break, highlight the row above where there is a page break and the ⊙ **Manual Page Break** button.

15.5 Understanding Reports

The **Reports** are found at the **Reports** tab. There are two types of reports:

- **Visual** which were introduced with Microsoft Project 2007 and

- **Reports** which have replaced the old style reports that have been available for many years and were unchanged since very early versions of Microsoft Project. These reports have a very significant and improved functionality and are the major enhancement in Microsoft Project 2013.

15.6 Visual Reports

Visual Reports allow:

- The creation of reports, such as Histograms and Cash Flows. These are generated from templates to be exported and displayed in Excel or Visio.

- The templates may be modified and new ones created.

To run a **Visual Report** select **Reports**, **Export** group, **Visual Reports** to open the **Visual Reports - Create Report** form.

- The term **Create Report** means that you are creating from a report template.

- In this form you may use an existing report template to create a report, create a new template, modify a template or create a new template.

- Templates are either used to display the report in Visio or Excel and the template files may be managed using the **Manage Template** option.

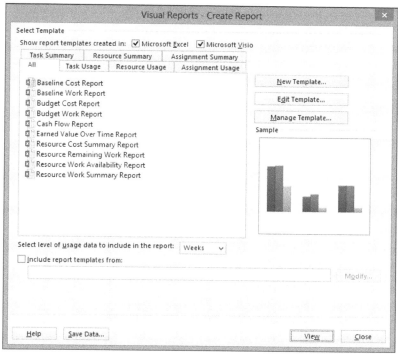

 The author found that Office 2010 or later had to be used for these reports to be exported to Excel.

15.7   Microsoft Project Reports

15.7.1   Understanding the Microsoft Project Reports

The **Reports** menu is new to Microsoft Project 2013 and has a significant number of options and formatting functions. This section will explain some of the basic functions but not go into detail on all the options. Most of the functions are very similar to other Microsoft software making it simple to learn how to create and format Reports.

Select **Reports**, **View Reports** group, to see a list of 8 available report menus. Clicking on the ▾ will list the available reports for each menu.

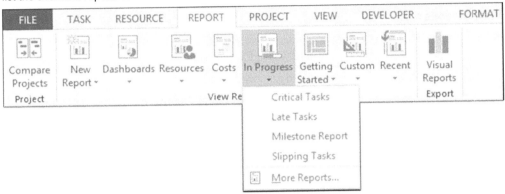

The basic principles of using these reports are:

- The reports are displayed in the upper pane, so it would be best if the window was not split before running a report.

- Reports are constructed with predefined and configurable components which may be added or removed, resized and formatted once placed on the report. The available components are:
  - ➢ **Images**
  - ➢ **Shapes**
  - ➢ **Chart**
  - ➢ **Table**
  - ➢ **Text Boxes**
- New Reports are automatically made part of the project file in the same way as creating a new Filter or Table, and is saved when the project is saved.

- **Organizer** may be used to copy Reports to other projects or templates in the same way as Calendars, Filters and Tables etc. and copied to and from other projects or templates.

- An existing **Report** is selected from the **Report**, **View Reports** menu and is presented in the upper window, where the Gantt Chart would normally be viewed.

- Components may be edited to show the required data. This is achieved by selecting the appropriate component with the mouse and then edited using editing menus which are displayed automatically after a report component is selected with the mouse.

- The report data is live and the Reports do not have to be "Run" as in the Visual Reports; they are just displayed and then printed using the normal print commands.

- Select the appropriate **View** to return back to the Gantt Chart or any other view you wish to return to using the normal menu commands such as the **Task**, **View** command.

15.7.2 Displaying and Formatting Microsoft Project Reports

To display and format a new Microsoft Project Report:

- Select the required Report from the **Reports**, **View Reports** menu and it will be displayed in the **Upper Pane**. It is best to un-split the window when displaying reports.

- To edit or format the report data:

  ➤ Select any component of the report with the mouse,

  ➤ The **Field List** form, which is normally automatically displayed on the right of the screen, may be used to change the data that is displayed,

  ➤ The **Field List** form may be displayed manually using the menu when it has been minimized,

  ➤ Many formatting options will be displayed on the **Ribbon** menu and are displayed under Ribbon tabs titled **Report Tools** and **Chart Tools**:

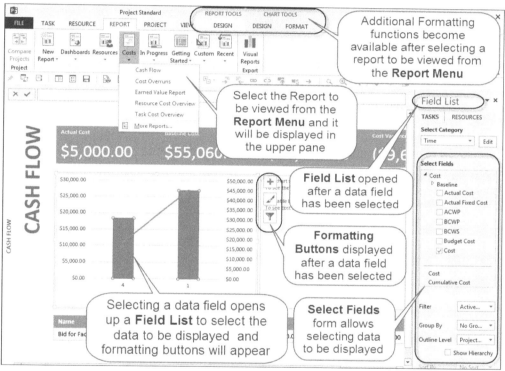

15.7.3 Creating New Microsoft Project Report

To create a new Microsoft Project Report:

- Select **Report**, **View Reports**, **New Report**,

- Select a template; the blank one will only present a report title,

- Enter the name of the report in the **Report Name** form and a new report will be displayed.

To add a component to a new Microsoft Project Report:

- The **Design** Ribbon menu command becomes available:

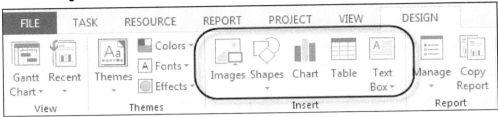

- You may select to insert one of the following components:
  - ➢ **Images** – Images may be placed on the report and could include a company logo for example.
  - ➢ **Shapes** – Shapes are the normal menu of Microsoft Office shapes that may be placed on the report and formatted in the normal way, these are normally shapes such as arrows, boxes etc.
  - ➢ **Chart** – A chart may be placed on a report and the clicking on a Chart will open a menu that allows the data to be selected and the method of data display, such as type of line or bar to be selected.
  - ➢ **Table** – A table may be placed on a report and the clicking on a Table will open a menu that allows the data to be selected and formatted.
  - ➢ **Text Boxes** – Text Boxes may be placed on the report and formatted in the same way as in other Microsoft Products. These would often be used for report headings.

To remove a component from a report, select the component and delete.

The sample report below was produced by inserting one of each of the Report components and formatting individually by clicking on the component and using the available menus.

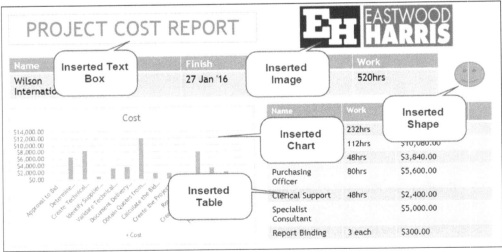

15.8   *Workshop 13 – Printing*

Background

We want to issue a report for comment by management.

Assignment

1.   We will baseline and progress the project in the next workshop, but we will require an unprogressed copy of your schedule for the resources workshops.

   ➢   Save your OzBuild project as **OzBuild With Resources**.

   ➢   Close the project and

   ➢   Reopen your **OzBuild** project.

2.   Apply the **Gantt Chart Inc Total and Neg Float** view.

3.   Select apply the **Entry** Table.

4.   Ensure there is no filter applied.

5.   Display only the Task ID, Indicators, Name, Duration, Start Date, Finish Date, Total Slack and Predecessors columns.

6.   Adjust the columns to the best fit.

7.   To fit all the tasks on one A4 or Letter size landscape page adjust the middle tier unit of timescale to **Months** with **Label** of **Jan '09** and bottom tier to **Weeks** with **Count:** of 1, **Label** of **26,2…** and **Size** of **150%**.

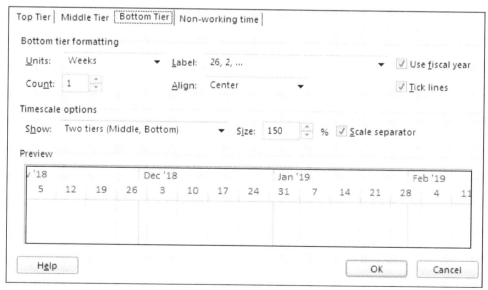

---

8. Check you have the following information into the **Project Properties** form:

9. Select **File**, **Print**:

➤ Select your printer

➤ Select Paper Size of A4 or Letter,

➤ Sometime it will try to print on two pages, with one being blank, in the **Page Setup**, **View** tab try unchecking the **Print blank pages,**

➤ In the **Page Setup**, **View** tab check the **Print all sheet columns,**

Close the form,

10. Select **OK** then back to **Print Preview**:

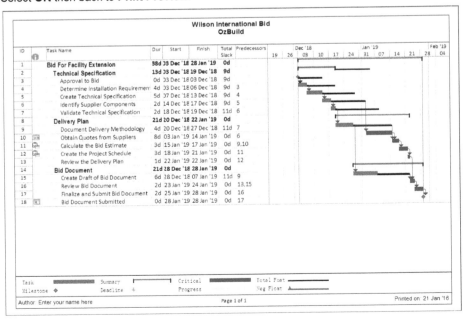

11. Save your **OzBuild p**roject.

16 TRACKING PROGRESS

The process of tracking progress is used after the plan has been completed, the Baseline set and the project is underway. Now the important phase of regular monitoring begins. Monitoring is important to help predict problems as early as possible, and thus minimize the impact of problems on the successful completion of the project.

Many schedules are updated in Microsoft by just entering the task % Complete and not adjusting the Actual dates and durations. The result of this process is an un-impacted progressed schedule which will visually indicate which tasks are ahead of time or behind time, but the current dates are always equal to the Baseline dates. This process results in a schedule that:

- Does not reforecast a revised project finish date,

- May not be used to calculate delays, extensions of time or acceleration,

- May result with completed work in the future and incomplete work in the past which is not logical.

This book will focus on the process of updating a schedule to create an impacted schedule which will reforecast a revised project finish date. The main steps for monitoring progress are:

- Set the **Baseline Dates**, also known as **Target Dates** in some software. These are the dates against which progress is compared.

- Approve the work to commence in accordance with the plan.

- Record or mark-up progress as of a specific date, the **Status Date**. This date is also known as the **Data Date**, **Current Date**, **Report Date**, **Update Date**, **Time Now** and **As-of-Date**.

- **Update** or **Status** the schedule with **Actual Start** and **Actual Finish** dates where applicable, and adjust the task **Actual Durations** and **Remaining Durations** and/or % **Complete**.

- Compare and **Report** actual progress against planned progress and revise the schedule, if required, to forecast future tasks and milestones.

- Implement any corrective action to bring the project back on schedule.

By the time you get to this phase you should have a schedule that compares your original plan with the current plan, showing where the project is ahead or behind. If you are behind, you should be able to use this schedule to plan appropriate remedial measures to bring the project back on target.

This chapter will cover the following topics:

Topic	Menu Command
- Setting, resetting and clearing the **Baseline**	Select **Project**, **Schedule** group, **Set Baseline**, **Set Baseline...**
- **Current Date** and **Status Date**	May be edited from the **Project**, **Properties** group, **Project Information** form.
- Automatically update the schedule	Select **Project**, **Status**, **Update Project**
- Move the **Incomplete Work** of an **In-progress** task into the future	Select **Project**, **Status**, **Update Project**
- Updating the Tasks	Select **Task**, **Schedule** group

16.1 Setting the Baseline

Setting the Baseline copies the following information into Baseline fields in the existing file:

- **Start** and **Finish** dates into fields titled **Baseline Start** and **Baseline Finish**.

- **Duration** into the **Baseline Duration**.

- Each resource's **Costs** and **Work** into **Baseline Costs** and **Baseline Work** (work is typically the number of hours) when resources are assigned to tasks.

 Unlike other scheduling products Microsoft Project does not take a complete copy of all the data and does not baseline constraints, relationships, float values and the critical path. Unlike P6, Microsoft Project does allow a Baseline for a partial project.

Once the Baseline is set and changes are made to the schedule, you will be able to compare your progress with your original plan and you will be able to see if you are ahead or behind schedule and by how much.

Some important points you should understand before setting the Baseline:

- The Baseline Dates should be established before you update the schedule for the first time and are usually subject to an official approval process.

- The Baseline Dates and Duration fields are not calculated fields, so they may be edited in columns and forms. Caution should be used before changing a Baseline Date or Duration, since it is the basis for all project deviation measurements.

- Summary tasks, scheduled Start, Finish, Duration, Costs and Work are recalculated as detail tasks are added or deleted. Setting the Baseline copies these values but the Baseline values are **NOT** recalculated when new tasks are created or existing tasks are moved to a different Summary task.

- Setting the Baseline Date does not store the logic, float/slack times, critical path or constraints.

- A Baseline is normally applied to all tasks irrespective of their Outline levels.

- New tasks do not have a baseline when created.

16.1.1 Setting Baseline Dates

To set the Baseline, select **Project**, **Schedule** group, **Baseline**, **Set Baseline...** to display the **Set Baseline** form.

- The **Set Baseline** option copies the **Start** and **Finish** dates and **Duration** into the **Baseline** fields.

- When resources and costs have been assigned then these are baselined at the same time.

- You may highlight some tasks before opening the **Set Baseline** form and click on the **Selected tasks** radio button to set the Baseline for the specified tasks only.

This operation will overwrite any previous Baseline settings.

The **Set Baseline** form default settings selects and baselines **All Tasks** and the task Start and Finish dates which are titled **Start/Finish**. This is the normal method of setting the Baseline.

After establishing the Baseline, you can input the progress data without fear of losing the original dates. This will also enable you to compare the progress with the original plan.

16.1.2 Clearing and Resetting the Baseline

You may clear some or all of the Baseline fields using the **Clear Baseline** form. This is opened by selecting **Project**, **Schedule** group, **Baseline**, **Clear Baseline plan**:

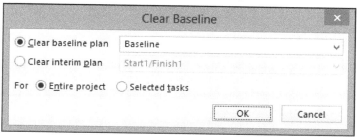

16.1.3 Displaying the Baseline Data

The Baseline dates data may be displayed by:

- Displaying the Baseline columns where the data may be edited, or
- Displaying the dates in a form such as the Resources Details Form. This will only display the **Baseline** data and not **Baseline 1** to **10**, or
- Show a baseline bar on the Bar Chart.

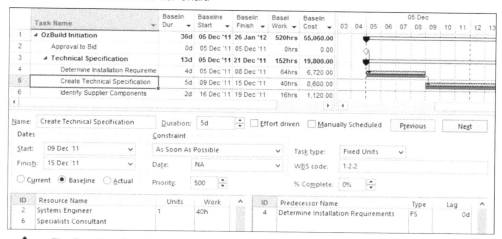

 The Baseline Start Dates, Finish Dates and Durations may be edited and are not linked by logic. Therefore, a change to a Baseline Duration will not affect either the Baseline Start Date or the Baseline Finish date.

16.2   Practical Methods of Recording Progress

Normally a project is updated once a week, fortnightly, or monthly. Very short projects could be updated daily or even by the shift or hour. As a guide, a project would typically be updated between 12 and 20 times in a project life cycle. Progress is recorded on or near the **Status Date** and the scheduler updates the schedule upon the receipt of the progress information.

The following information is typically recorded for each task when updating a project:

- The task Actual Start date and time if required,
- The number of days/hours remaining on the **Task** or when the task is expected to finish,
- The percentage complete, and
- If complete, the task Actual Finish date and time, if required.

A marked-up copy recording the progress of the current schedule is often produced prior to updating the data in Microsoft Project. Ideally, the mark-up should be prepared by a physical inspection of the work or by a person who intimately knows the work. It is good practice to keep this marked-up record for your own reference at a later date. Ensure that you note the date of the mark-up (i.e., the data date) and, if relevant, the time. It is important to ensure that each Actual Start and Actual Finish date be tied to independent evidence such as daily reports or photographs when a project is likely to be subject to dispute resolution.

Often an Updating Report or mark-up sheet, such as the one below, is distributed to the people responsible for marking up the project progress. The marked-up sheets are returned to the scheduler for data entry into the software system.

A View, such as the one below, could be created for recording progress:

- It may have a filter applied to display only tasks that are in-progress or due to start in the next few weeks.
- A manual page break could be placed at each responsible person's band, so when the schedule is printed each team member would have a personal listing of their tasks that are either in-progress or due to commence. This is particularly useful for large projects.

	Task Name	Duration	% Comp.	Start	Actual or Expected Start	Finish	Actual or Expected Finish
	⊿ **Responsibility: Angela Lowe - Purchasing**	**17d**	**0%**	**16 Dec**	**NA**	**12 Jan**	**NA**
6	Identify Supplier Components	2 days	0%	16 Dec	NA	19 Dec	NA
10	Obtain Quotes from Suppliers	8 days	0%	03 Jan	NA	12 Jan	NA
	⊿ **Responsibility: Carol Peterson - Bid Manager**	**19d**	**0%**	**30 Dec**	**NA**	**26 Jan**	**NA**
11	Calculate the Bid Estimate	3 days	0%	13 Jan	NA	16 Jan	NA
12	Create the Project Schedule	3 days	0%	17 Jan	NA	19 Jan	NA
13	Review the Delivery Plan	1 day	0%	20 Jan	NA	20 Jan	NA
15	Create Draft of Bid Document	7 days	0%	30 Dec	NA	10 Jan	NA
16	Review Bid Document	2 days	0%	23 Jan	NA	24 Jan	NA
17	Finalise and Submit Bid Document	2 days	0%	25 Jan	NA	26 Jan	NA
	⊿ **Responsibility: Scott Morrison - Engineering**	**17d**	**0%**	**05 Dec**	**NA**	**29 Dec**	**NA**
4	Determine Installation Requirements	4 days	0%	05 Dec	NA	08 Dec	NA
5	Create Technical Specification	5 days	0%	09 Dec	NA	15 Dec	NA

Other electronic methods, discussed next, may be employed to collect the data. Irrespective of the method used, the same data needs to be collected.

The above View has been created by:

- Renaming Text 1 as Responsibility using **Project**, **Properties** group, **Custom Fields** form
- Entering the responsible person in Text, and
- Grouping by Text 1, which has been renamed **Responsibility**.

There are several methods of collecting the project status:

- Physical inspection of the work,
- By talking to the responsible people,
- By sending a hard copy to each responsible person to mark up and return to the scheduler.
- By cutting and pasting the data from Microsoft Project into another tool, such as Excel, and emailing the data to each responsible person as an attachment.
- By giving the responsible party direct access to the schedule software to update it. This approach is not recommended, unless the project is broken into sub-projects. By using the sub-project method, only one person updates each part of the schedule.
- Microsoft Project Server, a companion Microsoft Project product that allows collaborative scheduling, and time sheeting could be implemented. This topic is beyond the scope of this book.
- By using a task-based timesheet system which would also collect the hours expended on each task, and potentially the estimated remaining work and duration required to complete the task.

Some projects involve a number of people. In such cases, it is important that procedures be written to ensure that the status information is collected:

- In a timely manner,
- Consistently,
- In a complete manner, and
- In a usable format.

 It is important for a scheduler to be aware that some people have great difficulty in comprehending a schedule. When there are a number of people with different skill levels in an organization, it will be necessary to provide more than one method of updating the data. You even may find that you have to sit down with some people to obtain the correct data, yet others are willing and comfortable to email you the information.

16.3   *Understanding Tracking Progress Concepts*

There are some terms and concepts used in Microsoft Project that must be understood before we update a project schedule:

16.3.1   **Task Lifecycle**

There are three stages of a task's lifecycle:

- **Not Started** – The **Early Start** and **Early Finish** dates are calculated from the logic, **Calendars**, **Constraints** and the **Task Duration**.
- **In-progress** – The task has an **Actual Start** but is not complete. Milestones may be in-progress in Microsoft Project but in reality they should either be Not Started or Complete.
- **Complete** – The task is in the past, the **Actual Start** and **Actual Finish** dates have been entered into Microsoft Project, and they override the logic and constraints.

16.3.2   Actual Start Date Assignment

This section will explain how Microsoft Project assigns the **Early Start** of an **In-progress** task.

- When an **Actual Start** is entered into the **Actual Start** field, this date overrides the **Early Start date**. The predecessor logic and start date constraints are ignored. **NOTE:** If **Split in-progress tasks** is checked in the **File**, **Options**, **Schedule** tab, **Scheduling options for this project:** section, then a successor may split a task when the successor is commenced before the predecessor finish date.

- For an un-started task, **Actual Start** date is set to equal the **Early Start** date when:
  - ➢ A **% Complete** between 1% and 100% is entered, or
  - ➢ An **Actual Duration** is entered, or
  - ➢ An **Actual Finish** is entered, or
  - ➢ An update schedule or task function is used.

- When a % Complete of 100% is entered, then an **Actual Start** date is set equal to the **Early Start** and an **Actual Finish** date equal to the **Early Finish** date are both set.

16.3.3   Calculation of Actual & Remaining Durations of an In-progress Task

- The **Actual Duration** is normally the worked duration of a task.
- The **Remaining Duration** is the un-worked duration of a task.
- **Duration = Actual Duration + Remaining Duration**. Before a task is commenced the **Actual Duration** is zero and the **Remaining Duration** equals the **Duration** assigned to the task.
- There is an in-built proportional link between **Duration**, **Actual Duration**, **Remaining Duration** and **% Complete**. It is not possible to unlink these fields (as in other scheduling software) and therefore not possible to enter the **Remaining Duration** independently of the **% Complete**. The following explains the relationship between these fields:
  - ➢ Change the **Duration**: the **Actual Duration** remains constant and the **% Complete** and the **Remaining Duration** changes proportionally.
  - ➢ Change the **% Complete**: the **Duration** remains constant and the **Actual Duration** and the **Remaining Duration** changes proportionally.
  - ➢ Change the **Actual Duration**: the **Duration** remains constant and the **% Complete** and the **Remaining Duration** changes proportionally.
  - ➢ Change the **Remaining Duration**: the **Actual Duration** remains constant and the **% Complete** and the **Duration** changes proportionally.

- When the **% Complete** is set to 100 or the **Remaining Duration** is set to zero, the **Actual Finish** date is set to the **Early Finish date** and the **Actual Duration is** set to the **Duration**.

The **Physical % Complete** column maybe displayed and updated when a task % Complete is required that is not related to the Durations. This leaves % complete as % time complete. The **Physical % Complete** field does not summarize to summary tasks and may be displayed on the bars.

The example below shows three tasks, the first un-started, the second in-progress and the third complete. You should observe:

- The relationship between the **Duration**, **Actual Duration**, **Remaining Duration** and **% Complete** in each of the tasks, and

- How the **Actual Start** and **Actual Finish** are set.

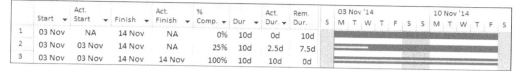

	Start ▾	Act. Start ▾	Finish ▾	Act. Finish ▾	% Comp. ▾	Dur ▾	Act. Dur. ▾	Rem. Dur.
1	03 Nov	NA	14 Nov	NA	0%	10d	0d	10d
2	03 Nov	03 Nov	14 Nov	NA	25%	10d	2.5d	7.5d
3	03 Nov	03 Nov	14 Nov	14 Nov	100%	10d	10d	0d

Unlike in Oracle Primavera software, it is not possible in Microsoft Project to unlink the **Remaining Duration** and **% Complete**. This may prove frustrating to some schedulers, as you will not be able use the **% Complete** column to represent the amount of deliverables complete when the task is not progressing at a linear rate. The Physical % Complete could be used to represent the amount of deliverables complete when the task is not progressing at a linear rate.

Furthermore, Microsoft Project calculates the Finish Date of a task from the Actual Start plus the Duration. It ignores the Remaining Duration, the Current Date and Status Date when calculating the Finish Date of a task. Oracle Primavera users will find this a difficult concept to understand but these concepts must be understood to use Microsoft Project effectively.

16.3.4 Calculating the Early Finish Date of an In-Progress Task

Retained Logic and Progress Override are not terms used by Microsoft Project documentation. These terms are used by other Oracle Primavera software and are used here to help clarify how Microsoft Project performs its calculations. In the example below, there are two tasks with a Finish-to-Start relationship:

	Act. Start ▾	Act. Finish ▾	% Comp. ▾
1	NA	NA	0%
2	NA	NA	0%

There are two options for calculating the finish date of the successor when the successor task starts before the predecessor task is finished:

- **Retained Logic.** In the example below, the logic relationship is maintained between the predecessor and successor for the un-worked portion of the Task, the Remaining Duration, and continues after the predecessor has finished.

	Act. Start ▾	Act. Finish ▾	% Comp. ▾
1	05 Nov	NA	50%
2	06 Nov	NA	10%

This option will operate when:

➢ The Project Option **Split in-progress tasks** is checked, and

➢ There is an **Actual Start**, and

➢ A **% Complete** between 1% and 99% is assigned to the successor task.

- **Progress Override**. In the example below, the Finish-to-Start relationship between the predecessor and successor is disregarded, and the un-worked portion of the Task, the Remaining Duration, continues before the predecessor has finished:

Progress Override option will operate when:

➢ The Project Option **Split in-progress tasks IS NOT** checked, or

➢ The task has an **Actual Start** and 0% Complete and Project Option **Split in-progress tasks IS** checked.

  Be aware that this function calculates differently to earlier versions of Microsoft Project and will split a successor task when an actual start is assigned to a successor that is earlier than the Start date and the % Complete is zero.

This function may result in some problems in reporting a schedule when the **Split in-progress tasks** option is used. The two examples below are from the same schedule, both with the **Split in-progress tasks** option checked but the Actual date and % Complete were entered in a different order so different results are obtained:

➢ % Complete entered first

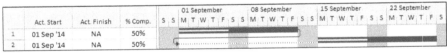

➢ Actual Start entered first

NOTE: You may wish to experiment with a Finish-to-Finish lag if these options do not give a satisfactory result.

You may wish to turn off **Auto Calculation** in the **File**, **Options**, **Schedule**, **Calculation** section before assigning an **Actual Start** as this will create a split at the start of the task when turned on and an Actual Start applied You will then need to press F9 to reschedule after updating the tasks.

16.3.5   Updating Completed tasks

Normally completed tasks are updated with actual dates that match when the task started and finished. To update a completed task:

- Enter the Actual Start date, the date the task actually started and then either:

  ➢ Enter the Actual Finish Date and then the % Complete will be set to 100% or

  ➢ Adjust the Remaining Duration until the task Finish Date is the same as when the task actually finished, then enter 100%.

- Alternatively enter 100% and adjust the Actual Start and then the Actual Finish.

16.3.6 Summary Bars Progress Calculation

Summary bars are not normally updated by entering the **Actual Start** date, **Actual Finish** date, or **% Complete** against them. This is possible in Microsoft Project and is covered later in the section titled **Marking Up Summary Tasks**, see para 24.9.6. It is recommended that this process is not used.

This status information is usually entered against the detail tasks, so the summary tasks inherit the status data from the detail tasks.

- An **Actual Start** is assigned against a summary task when any detail task has an **Actual Start**.
- A summary task's **% Complete** is calculated from the total of all the detail tasks' **Actual Durations** divided by the total of all the detail tasks' **Durations**. This may be misleading when Milestones have been marked complete and no progress is calculated on the summary bar.
- A summary task's **Actual** and **Remaining Durations** are calculated from the **Duration** and **% Complete**.
- An **Actual Finish** is assigned against a summary task when all detail tasks have an **Actual Finish**.

16.3.7 Understanding the Current Date, Status Date & Update Project Date

Microsoft Project has two project Data Date fields that may be displayed as vertical lines on the schedule. These dates may be edited from the **Project, Properties** group, **Project Information** form:

- **Current Date** – This date is set to the computer's date each time a project file is opened. It is used for calculating **Earned Value** data when a **Status Date** has not been set. By default this is set to **Default start time:** in the **File, Options, Schedule** tab, **Calendar options for this project:** section. This is not generally used.
- **Status Date** – This field is blank by default with a value of **NA**. When this date is set, it will not change when the project is saved and reopened at a later date. When set, this date overrides the **Current Date** for calculating **Earned Value** data. After this date has been set then it may be displayed as a gridline and you may remove the date by typing **NA** into the field. By default this is set to **Default end time** in the **Options** form.

 NOTE: The **Current Date** is set to the **Default start time** and the **Status Date** is set to **Default end time**, so when both are set to the same date and displayed as grid lines, they will both be displayed, one at the start and one at the end of the day. See the picture below:

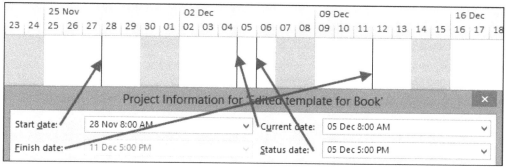

 It is recommended that the **Status Date** be set and displayed as a vertical line on a progressed schedule and that the **Current Date** not be displayed, as the **Current Date** represents the date today and does not normally represent any scheduling significance.

The **Update Project Date** may also influence how Microsoft Project calculates the end date of some tasks. This date may not be displayed as a vertical line on the screen but may be used in conjunction with the **Reschedule uncompleted work to start after:** function covered later in this chapter.

NEITHER the **Current Date** nor the **Status Date** is used to calculate the **Early Finish** of an **In-progress** task of a schedule using **F9** or **Automatic Scheduling**, unless one of the following two functions is used:

- The **Project**, **Status** group, **Update Project...**, **Reschedule incomplete to start after:** date, or
- The **Status Date** may be used to move the start of incomplete and the finish of completed parts of a task back or forward to the **Status Date** with the function **Status Date Calculation Options** (new to Microsoft Project 2002). This option will be discussed later in this chapter in para 0. It has restrictions that could make the function difficult to use on a project.

Project, **Status** group has the **Status Date** function that displays the **Current Status Date**, it will open the **Status Date** form where the status date may also be edited.

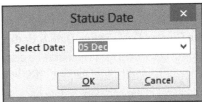

 Ideally, scheduling software has one current **Data Date** and the function of it is to:

- Separate the completed parts of tasks from incomplete parts of tasks,
- Calculate or record all costs and hours to date before the data date, and to forecast costs and hours to go after the data date.
- Calculate the **Finish Date** of an in-progress task from the **Data Date** plus the **Remaining Duration** over the **Task Calendar**.

It is relatively easy to be in a situation where you have complete or in-progress tasks with start dates later than the data date, and/or incomplete or un-started tasks with a finish date earlier than the data date. This is an unrealistic situation, which is more difficult to achieve in other scheduling software packages. Care should be taken to avoid this situation and checks made after the schedule has been updated.

16.4 Updating the Schedule

The next stage is to update the schedule by entering the mark-up information against each task.

When dealing with large schedules, it is normal to create a look-ahead schedule with a filter to display incomplete and un-started tasks commencing in the near future only.

Microsoft Project provides several methods of updating the schedule:

- **Update Project** – This is an automated process that assumes all tasks have progressed as planned. After updating a project, you may adjust the dates and percent complete to the recorded progress.
- **Update Tasks** – This function is used to update tasks one at a time.
- Update tasks using the **Task** or **Task Details** form.

- Update tasks by displaying the appropriate tracking columns by:
  - ➤ Selecting the **Tracking** table, or
  - ➤ Creating your own table, or
  - ➤ Inserting the required columns in an existing table.

Each of these methods is discussed in the following sections. Then we will discuss the following two functions which are designed to assist you in moving tasks to their logical places in relation to the **Status Date**:

- Move incomplete work into the future, using the **Reschedule uncompleted work to start after: function**, see para 0 and

- Positioning tasks in a logical place in relation to the Status Date using the **Status Date Calculation Options** function, see para 24.9.4.

16.4.1 Move Project

This function was new to Microsoft Project 2010.

Project, **Schedule** group has the **Move Project** button which opens the **Move Project** form and allows the user to change the **Project Start Date**. When the **Move deadlines** check box is checked any Deadline Dates and constraints are also moved based on a 7-day per week calendar so moving a project one day will move a **Deadline Date** on a Friday to a Saturday:

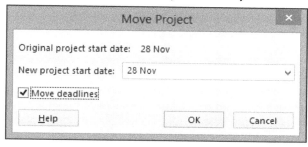

 Baselines are not moved with this function

16.4.2 Using Update Project

Microsoft Project has a function titled **Update Project** for updating a project as if it had progressed according to plan. This function sets **Actual Start** and **Actual Finish** dates, % **Complete** and **Renaming Durations** in proportion to a user-assigned date.

Add a button to the **Quick Access Toolbar** or select **Project**, **Status** group, **Update Project…** to open the **Update Project** form::

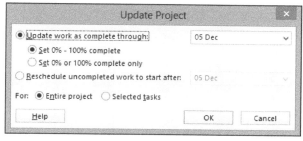

The two options under **Update work as complete through:** which apply to in-progress tasks only:

- **Set 0% – 100 % Complete**:
  - ➤ This option sets the **Status Date** to the date you have nominated as the update date. This is displayed as the vertical line in the Gantt Chart below create using the **Format Gridlines** function,
  - ➤ Sets the **Actual Start** to the **Start**, and
  - ➤ Sets the **% Complete** and **Actual Duration** in proportion to the amount of time worked for any in-progress tasks.

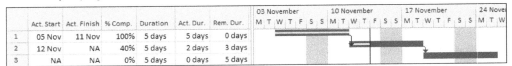

- **Set 0% or 100 % Complete only**. This option sets:
  - ➤ The **Actual Start** to the **Early Start** but leaves the **% Complete** and **Actual Duration** at zero.
  - ➤ The **% Complete** is set to 100% only when the task is complete.
  - ➤ Does not set the **Status Date** (or reset if it has been set in a previous update) to the date you have chosen as the update date. There is no vertical line representing the **Status Date** in the picture below:

You may use the **Selected tasks** option when you highlight the tasks to be progressed before opening the **Update Project** form. Unselected tasks will not be progressed, but these tasks may be scheduled to occur after the **Status Date** by using the **Reschedule uncompleted work to start after:** function. This is covered later in this section.

You may not reverse progress with this option as one is able to do with SureTrak and Asta Powerproject.

16.4.3 Update Tasks Form

Microsoft Project has a function that may be used for updating tasks one at a time or may be used for updating selected tasks with the same information. For example, there may be several tasks with the same Actual Start or % Complete:

- Select one or more tasks that you want to update with the same information such as the same Actual Start date.

- Select **Task**, **Schedule** group, **Mark on Track**, ⊟☑ **Update Tasks...** to open the **Update Tasks** form:

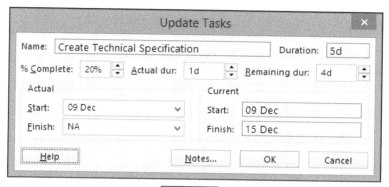

- Enter the required data and click on the [OK] button to close the form.

16.4.4 Updating Tasks Using the Task Information Form

You may use the **Task Information** form to update progress:

- Open the **Task** form by double-clicking on a task,
- Then the Percent complete and the dates may be edited.

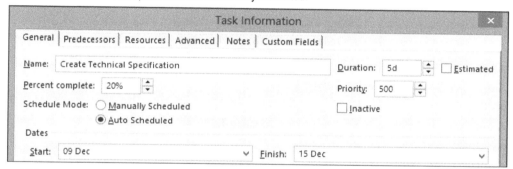

  The **Task Information** form should not be used when the Actual and Remaining durations are being adjusted to update a project.

  Using the **Update Tasks** form or the **Task Information** form are cumbersome methods for updating a large number of individual tasks since the form has to be closed after each task has been updated, the cursor moved to the next task and the form reopened.

If you have a large number of tasks to be updated then it is quicker to use:

- Columns, or
- A Detailed form in the bottom pane, or
- The Task, Schedule group buttons.

16.4.5   Updating Tasks Using the Task Details Form

- To open the **Task Details** form, open the dual-pane view using the right-click **Show Split** command,
- Now open the **Task Details Form** by right-clicking in the band on the left-hand side of the screen,
- Select <u>More Views</u>... to open the **More Views** form,
- Select **Task Details Form**.

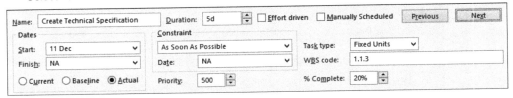

- Updating a task:
  - ➢ Click on the **Actual** radio button,
  - ➢ Enter the Actual Start date,
  - ➢ Enter a **% Complete**. An **Actual Start** date will be set by Microsoft Project when a % Complete is entered and may be edited from the <u>Start:</u> drop-down box,
  - ➢ The **Duration** may be edited in order to calculate a new finish date. You will note that the % Complete will change when durations are edited.
  - ➢ When a date is typed into the **Fini<u>s</u>h:** drop-down box, a **Finish Constraint** may be set without warning when the task is not 100% complete. These may be edited from the **Task Information** form's **Advanced** tab.

16.4.6   Updating Tasks Using Columns

An efficient method of updating tasks is by displaying the data in columns. This may be achieved by:

- Applying the **Tracking Gantt** view and the **Tracking Table**, or
- Creating your own table, or
- Inserting the required columns.

The tracking Table may not exist if you are working on a schedule that has had this table deleted or if you have a non-standard install of Microsoft Project in which this table has been deleted from the Global.mpt. If you do not use the tracking table, be sure to add the following columns: **Actual Start**, **Actual Finish**, **% Complete**, **Actual Duration**, and **Remaining Duration**.

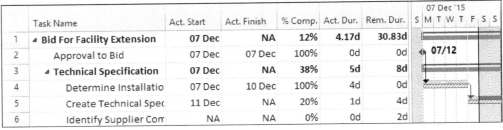

The status data may now be entered directly into the columns.

16.4.7 Reschedule Uncompleted Work To Start After

There is a feature available in the **Update Project** form titled **Reschedule uncompleted work to start after:** which will schedule:

- The **Incomplete Work** of an **In-progress** task to start on a specific date in the future, and

- Move **Un-started tasks** to start on the Status Date by assigning a constraint as part of the process. This is a very different method of calculation to most other products like P6 and Asta Powerproject, which do not need a constraint to keep un-started tasks in the future.

If you want to apply this operation to some tasks, then these options should be selected first.

- Add a button to the **Quick Access Toolbar** or select **File**, **Options**, **Schedule** tab and ensure the **Split in-progress tasks** option is checked. If this option is not checked, then this function will not operate.

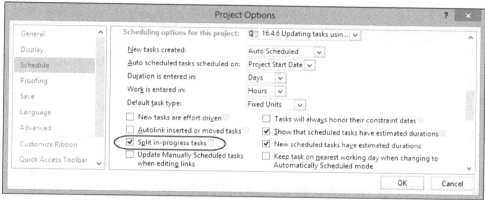

- Add a button to the Quick Access Toolbar or select **Project**, **Status** group, **Update Project...** to open the **Update Project** form:

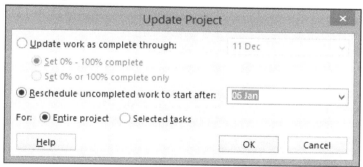

Click on the **Reschedule uncompleted work to start after:** radio button.

- In the drop-down box to the right, specify the date after which incomplete work should commence.

- Then click on the [OK] button.

In the examples below, the **Incomplete Work** has been moved after 10 September:

- The task has been split in the picture below, the splits are displayed and the **Remaining Duration** is dependent on the Task Type, **Fixed Duration** tasks have a longer duration than **Fixed Units** and **Fixed Work. Fixed Duration** tasks calculate their duration from the task start to the task finish, whereas **Fixed Units** and **Fixed Work** tasks add the portion of tasks together:

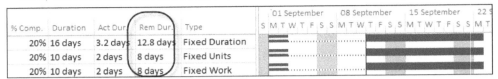

- The splits may be hidden using the **Format, Format** group **Layout** form but at the time of writing the book this function did not work and gave an error message. This may be fixed in updates to the product.

A task may be split may be hidden in the Gantt Chart by un-checking the **Show bar splits** in the **Format, Format** group **Layout** form.

When a project is updated using the **Update Project** function, the **Status Date DOES NOT** change to equal the **Project Update Date**. You may end up with two dates that represent the **Data Date** - the **Update Project** date and the **Status Date**, which may both reflect different dates. The picture below shows the Status Date in the black vertical line at 10 September and the **Update Project** date set at 15 September, which has split the second task. In this situation the Status Date should be set manually.

For tasks that are behind schedule, you may use the **Update Project** function to reschedule them after the **Update Project** date. This date will effectively become the **Data Date**. However, it is not possible to show the **Update Project** date as a vertical line on the bar chart. You may set and display the **Status Date** at the same date so incomplete work is all scheduled after a line displayed on the Gantt Chart.

It is recommended when you use the **Update Project** function that you should open the **Project Information** form and set the **Status Date** to the same date as the **Update Project** date. This enables you to display a **Data Date** on the bar chart in the correct place.

The following two pictures show how the software will assign a constraint to an un-started task to hold it out past the date used with the **Reschedule uncompleted work to start after:**

Before Applying:

	ⓘ	% Complete	Duration	Remaining Duration	02 December							09 December						
					S	M	T	W	T	F	S	S	M	T	W	T	F	S	S
1		0%	4d	4d															
2		50%	4d	2d															

After Applying with a date of 06 Dec, note the constraint date:

	ⓘ	% Complete	Duration	Remaining Duration	02 December S M T W T F S	09 December S M T W T F S S
1		0%	4d	4d		▬▬▬▬
2	⊞	This task has a 'Start No Earlier Than' constraint on 06 Dec.			▬▬▬▬	▬▬▬

16.4.8 Updating Tasks Using Task, Schedule Functions

The **Task**, **Schedule** group has the following functions. It has replaced the original **Tracking Toolbar** from earlier versions of Microsoft Project

- The `0% 25% 50% 75% 100%` buttons will set the Percentage Complete as indicated. These will set an Actual Start Date and may be used to remove progress from a task.

- The **Mark on Track** button will progress the task to the Status Date as if it has progressed according to plan.

16.4.9 Reschedule Work Function

The **Reschedule Work** button that may be added to the **Quick Access Toolbar,** is very useful to reschedule work of selected in-progress tasks that have incomplete work in the past.

The **Split In Progress** option must be enabled in the **File**, **Options**, **Schedule** and the two pictures below show the effect of applying this option with the tasks 1, 2 and 3 selected before applying the function:

- Before, with Task 1, 2 and 3 selected:

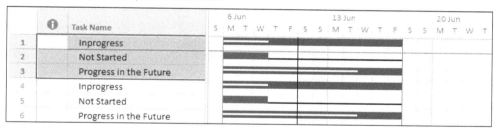

- After, Task 1 split, Task 2 moved to the future but Task 3 with complete work in the future is unaffected and needs manual intervention to ensure complete work is in the past. Task 4, 5 and 6 are also unaffected:

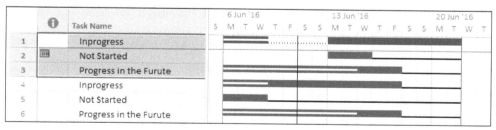

16.4.10 Updating Milestones Issues

It is very important when updating a Milestone that the **Actual Finish** date is **NEVER** entered as this will convert the Milestone to a Task.

To update a Milestone you should; enter the **Actual Start** date and then 100%.

The two pictures below show what happens to a Milestone when an **Actual Start** and **Actual Finish** date is entered:

- Before Updating:

Task Name	Act. Start	Act. Finish	% Comp.	F	S	S	M	T	W
Start Milestone	NA	NA	0%				◆		

- After Updating by entering **Actual Start** and **Actual Finish** date:

Task Name	Act. Start	Act. Finish	% Comp.	F	S	S	M	T	W
Start Milestone	13 Dec '16	13 Dec '16	100%				▬		

- After Updating by entering **Actual Start** and **100%**:

Task Name	Act. Start	Act. Finish	% Comp.	F	S	S	M	T	W
Start Milestone	13 Dec '16	13 Dec '16	100%				◆		

16.4.11 Physical % Complete

The **% Complete** field is the relationship between the **Actual Duration** and the **Duration;** whereas the **Physical % Complete** column has no link to any other data fields, unless it is selected in **File, Options**, **Advanced** to calculate **Earned Value.**

The **Physical % Complete** column may be displayed and updated when a task % Complete is required that is not linked to Durations. This is similar to the P6 Physical % Complete.

The Physical % Complete field does not summarize to summary tasks.

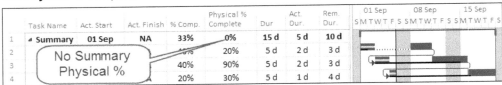

The Physical % Complete bar may be displayed by using the formatting shown in the picture below:

This publication is only sold as a bound book, no parts may be reproduced by any means, e.g. electronic, video or print.

© **Eastwood Harris Pty Ltd**   214

16.5 Simple Procedure for Updating a Schedule

For those people who require just one simple method of updating a schedule utilizing the **Update Project** function then the following process should be considered. It may not suit all situations especially when a project is way off plan. It is ideally suited to a situation when the plan is being closely followed and only minor adjustments are required to the actual dates and durations:

- Ensure that everyone on the project team is aware of the reporting cycle, the updating procedure and review process.

- Collect accurate and complete status information relative to the **Status Date**.

- Apply a suitable view and table such as the **Tracking** layout and **Tracking** table.

- Set the Baseline by selecting **Project**, **Schedule** group, **Set Baseline**, **Set Baseline....**

- Display the Baseline bars by selecting **Format**, **Bar Styles** group and selecting **Baseline**.

- Display the **Status Date Gridline**, select **Format**, **Format** group, **Gridlines**, **Gridlines....**

- Select **Project**, **Status** group, **Update Project** to open the **Update Project** form and select **Set 0% – 100 % Complete**, set the date in the form to the new **Status Date**.

- The project should be updated as if it has progressed exactly as planned and the **Status Date** should now be displayed in the bar chart.

- Now adjust the task dates by dragging the bars or entering the dates in the appropriate column; the order that the actions take place is important:

  ➤ **Complete** tasks should have the **Actual Start** and then the **Actual Finish** dates adjusted to when the task actually started and actually finished.

  ➤ **In-progress tasks** should have the **Actual Start** adjusted first, then the task bar dragged or **Duration** adjusted so the **Finish Date** is where it is expected to finish and finally the % Complete adjusted. The **Mark on Track** button is good for this purpose.

  ➤ **Un-started** tasks should have their logic and durations revised.

- Add any scope changes to the schedule.

- Un-started tasks should have their logic and durations revised.

- Save the project with a new file name and save for future reference.

> The Eastwood Harris template found at www.eh.com.au has much of this formatting completed and the Tracking Table has an additional column showing what is required to do to ensure the tasks are correctly updated.

16.6 Procedure for Detailed Updating

This procedure is suited to people who wish to update a schedule properly and make sure the Actual dates and durations are correct. It has small but important differences to the previous process:

- Ensure that everyone on the project team is aware of the reporting cycle, the updating procedure and review process.

- Collect accurate and complete status information relative to the **Status Date**.

- Apply a suitable layout and table such as the **Tracking** layout and **Tracking** table.

- Set the **Baseline** by selecting **Project**, **Schedule** group, **Set Baseline**, **Set Baseline....**

- Display the Baseline bars by selecting **Format**, **Bar Styles** group and selecting **Baseline**.

- Display the **Variance** columns; the **Finish Variance** is always a popular column to display.

- Display the **Status Date Gridline** and hide the **Current Date** gridlines by selecting **Format**, **Format** group, **Gridlines**, **Gridlines....**

- Now enter the task status for each task one at a time by entering the information in the appropriate column.

- The order in which the data entry take place is important as different results will be obtained when data is entered in a different order:

  ➢ **Complete tasks** should have the **Actual Start** and then the **Actual Finish** dates adjusted, in this order, to the date that the task actually started and actually finished. If you adjust the Finish date first, then the Start date, you will have to readjust the Finish date again.

  ➢ **Completed Milestone**s will be changed to a Task if an Actual Finish date is entered, so only enter an Actual Start and 100% when a Milestone is complete,

  ➢ **In-progress tasks** should have the **Actual Start** entered first, then there are two options for updating the task:

    1. The task bar may be dragged or Duration adjusted so the finish date is where it is estimated to finish and the **% Complete** may be adjusted manually by typing in the value or dragging the % Complete in the Gantt Chart, or use the ⇥ **Mark on Track** button.

    2. When you have been provided with tasks' Remaining Durations or Expected finish dates. The Actual Duration should be entered so that progress is up to the **Status Date** and then the Remaining Duration may then be entered and the Actual Duration will not change.

Tasks that are behind schedule may be split with the ⊞⇥ **Reschedule Work** button. This may be added to the **Quick Access Toolbar** or **Ribbon** or selecting **Project**, **Status** group, **Update Project...** to open the **Update Project** form and clicking on the **Reschedule uncompleted work to start after:** radio button.

Make sure that the **File**, **Options**, **Schedule** tab, **Split in-progress tasks** box is checked and you may wish to turn off **Auto Calculation** in the **File**, **Options**, **Schedule**, **Calculation** section before assigning an **Actual Start** as this will create a split at the start of the task when turned on and an Actual Start applied. You will then need to press F9 to reschedule after updating the tasks.

- Un-started tasks should be in the future and have their logic and durations revised.

- Add new activities to reflect scope changes or adding further detail to the schedule.

- Save the project with a new file name and save for future reference.

The Eastwood Harris template found at www.eh.com.au has much of this formatting completed and the Tracking Table has an additional column showing what is required to do to ensure the tasks are correctly updated

16.7 Comparing Progress with Baseline

There will normally be changes to the schedule dates and more often than not these will be delays. The full extent of the change is not apparent without having a Baseline bar to compare with the updated schedule.

To display the **Baseline Bar** in the **Gantt Chart** you may use any of the functions covered in the **FORMATTING THE DISPLAY** chapter,

- The **Format**, **Bar Styles** group, **Baseline** button, or
- The **Format**, **Bar Styles** group, **Format**, **Bar Styles** function, or
- You may use the **Gantt Chart Wizard** as follows if your project was created in Microsoft Project 2007 and earlier. This is because the Microsoft Project 2013 and 2016 formatting is not compatible with Microsoft Project 2007 and earlier versions:

  ➢ Add the **Gantt Chart Wizard button** to your Quick Access Toolbar or Ribbon,

  ➢ Click on the **Gantt Chart Wizard button** to open the Gantt Chart Wizard.

  ➢ Select the **Baseline** or **Custom Gantt Chart** option. This will display both the current schedule and the baseline.

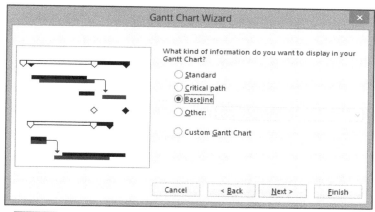

  ➢ Hit the **Next >** button and follow the remainder of the instructions to complete the formatting. You will be given options for applying text and relationships in the Gantt Chart.

> ⚠ This wizard will overwrite any customized formatting you have made using the **Format**, **Bar Styles** group options.
>
> The author found the bar formatting with the **Gantt Chart Wizard** is not compatible with the other formatting options provided with the new Microsoft Project 2013 **Format**, **Bar Styles** options. This is because the **Gantt Chart Wizard** puts the Baseline bar in the lower half of the bar, and the **Format**, **Bar Styles** group options puts the Baseline Bar in the upper half of the bar. Therefore, the **Gantt Chart Wizard** should only be used with files created in Microsoft project 2007 and earlier.

The Start and Finish Date variances are available by displaying the **Start Variance** and **Finish Variance** columns. These variance columns compare the **Baseline** dates with the current schedule dates. Variance columns and are not available for **Baseline 1** to **10 dates**, only the **Baseline** dates.

A correctly updated schedule should have, as the picture below:

- Completed tasks in the past,

- In-progress tasks should span the Status Date, with the Actual Duration in the past and Remaining Duration in the future,

- Un-started tasks in the Future:

	Task Name	Act. Start	Act. Finish	% Comp.	Dur	Act. Dur.	Rem. Dur.	13 October / 20 October / 27 Octo
1	A	13 Oct '14	17 Oct '14	100%	5d	5d	0d	
2	B	16 Oct '14	NA	60%	5d	3d	2d	
3	C	NA	NA	0%	5d	0d	5d	

An incorrectly updated schedule is as the picture below:

	Task Name	Act. Start	Act. Finish	% Comp.	Dur	Act. Dur.	Rem. Dur.	13 October / 20 October / 27 Octo
1	A	13 Oct '14	NA	80%	5d	4d	1d	
2	B	16 Oct '14	NA	40%	5d	2d	3d	
3	C	23 Oct '14	NA	40%	5d	2d	3d	

- This method is very popular with building companies,

- No Baseline is required and only the % Completes are entered,

- This method gives an indication of how progress is going but may not be used to manage a project or claim extensions of time or accelleration.

To correctly update the project above it should look something like the picture below:

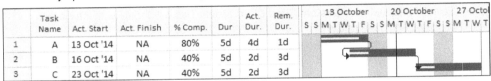

	Task Name	Act. Start	Act. Finish	% Comp.	Dur	Act. Dur.	Rem. Dur.	13 Oct / 20 Oct / 27 Oct
1	A	13 Oct '14	NA	80%	5 d	4 d	1 d	
2	B	16 Oct '14	NA	40%	5 d	2 d	3 d	
3	C	17 Oct '14	NA	40%	5 d	2 d	3 d	

16.8   In-Progress Schedule Check List

The following check list may be used to check an in-progress schedule.

Complete Tasks

- All tasks have an Actual Start and an Actual Finish before the Status Date.

In-Progress Tasks

- Actual Start dates should all be in the past and Early Finish dates in the future.
- Check tasks with Constraints, are the constraints still valid?
- Do all tasks have Finish successors?
- Are there any tasks with progress greater than planned and require the duration shortened?
- Are there any tasks that have progressed slower than planned and require the duration lengthened or split.

Not Started Tasks

- All Start dates should be in the future.
- Check that all the constraints on these tasks are still valid.
- All tasks should have Finish successors except the Project Finish Milestone(s).

Open Ends and Total Float

- Confirm that all tasks have successors and review float. Tasks with excessive float should be assigned dummy successors or delayed if they are not scheduled in a realistic timeframe with sequencing logic or Early Start constraints.

Critical and Near Critical Path

- Check the Critical Path is realistic and aligned with what project personnel consider critical.

Scope Changes

- Ensure that all project changes have been reflected in the schedule.
- It is good practice to add new activities for new work and cross reference the task to the change documentation using the Notes or a Text column.

Performance

- Has the performance of complete and in-progress tasks been reflected in un-started tasks, thus revising the un-started work durations based on achievements to date.

Baseline Comparison

- Review the Baseline dates with the current schedule and confirm that any delays are legitimate.
- Have there been many changes and delays and therefore should the schedule be re-baselined?

16.9   Corrective Action

There are two courses of action available with date slippage:

- The first is to accept the slippage. This is rarely acceptable, but it is the easiest answer.
- The second is to examine the schedule and evaluate how you could improve the end date.

Solutions to return the project to its original completion date must be cleared with the person/people responsible for the project, since they can have the most impact on the work.

Suggested techniques to bring the project back on track include:

- Reducing the durations of tasks on, or almost on, the Critical Path. When tasks have resources, increasing the number of resources working on the tasks may reduce duration. Changing longer tasks is often more achievable than changing the length of short duration tasks.
- Providing more time by changing calendars, say from a five-day to a six-day calendar, so that tasks are being worked on for more days per week or providing more hours per week.
- Changing task relationships so tasks take place concurrently. This may increase the peak resource requirements and this issue needs to be taken into account. This may be achieved by:
  - ➢ Introducing negative lags to Finish-to-Start relationships which maintains a Closed Network, or
  - ➢ Changing Finish-to-Start relationships to Start-to-Start which may result in open ends which will have to be closed by adding a Finish successor.
- Change the method of execution, for example moving work off-site.
- Reducing the project scope and hence deleting tasks.

16.10 Workshop 14 - Updating the Schedule and Baseline Comparison

Background

At the end of the first week you have to update the schedule and report progress and slippage.

Assignment

Open your **OzBuild Bid** project file and complete the following steps:

1. Ensure you have saved a copy of your project as **OzBuild With Resources** for use in the resources workshop and reopen your **OzBuild Bid** project.

2. Apply the **Gantt Chart** view.

3. Set the Baseline for all the tasks on your project using the **Project, Schedule** group, **Set Baseline, Set Baseline...** command and use the defaults.

4. Display the baseline bar by selecting **Format**, **Bar Styles** group **Baseline**.

5. Apply the **Tracking** table.

6. You will see a **Status Check** column, which is part of the Eastwood Harris template to assist in updating a schedule properly.

   ➢ This column **ONLY** works when the **Status Date** has been set, and

   ➢ **DOES NOT** calculate correctly on **Summary Tasks**; ignore Summary Task comments.

7. Set the timescale as per the picture below, with week and months.

8. Hide the **Physical % Complete**, **Actual Costs** and **Actual Work** columns.

9. Add the **Finish Variance** column on the right.

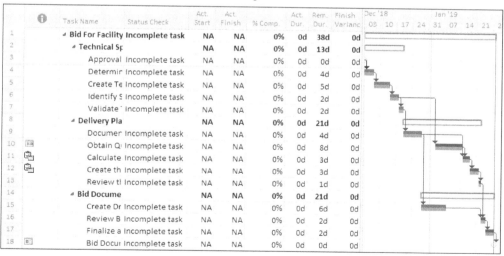

NOTE: The **Status Date** gridline will not display until the **Status Date** has been set.

10. Use **Project**, **Status** group, **Update Project...** to update progress to 07 Dec 18.
    NOTE: The **Status Date** time will be set at 5:00pm or 17:00 hrs on the Friday.

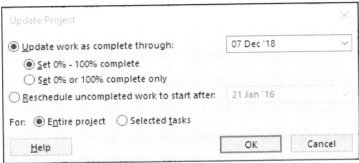

11. Check the **Status date** in the **Project**, **Properties** group, **Project Information** form; it should be 07 Dec 18 and the black **Status Date** line should now be displayed.

12. Expand the time scale to days using the 🔍 **Zoom** button, you also may need to adjust the **Size** in the **Timescale** form.

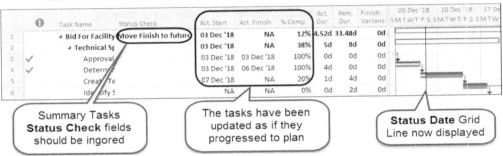

13. Update the following tasks using the table below to represent when the tasks were actually started and finished.

14. When you update tasks, you should enter the data in the exact order below, against the detail tasks only.

   • **Complete Milestones** should have the Actual Start and then 100 % assigned to the % Complete. Do not enter an Actual Finish against Milestones as they will change to tasks. Task 3 will just need the Actual Start change to 4 Dec 18 and 100% assigned.

   • **Complete Tasks** should have Actual Start and then Actual Finish updated. Task 4 will need both the Actual Start and Actual Finish adjusted.

   • **In-progress Tasks** should have Actual Start entered first, then Actual Duration so the Actual Duration meets the **Status Date** and then the Remaining Duration. There are other options outlined in this book for updating in-progress activities. Task 5 requires the Actual Start, Actual Duration and then Remaining Duration adjusted.

	Task Name	Act. Start	Act. Finish	Act. Dur.	Rem. Dur.
3	Approval to Bid	04 Dec '18	04 Dec '18	0d	0d
4	Determine Installation Requirements	04 Dec '18	06 Dec '18	3d	0d
5	Create Technical Specification	06 Dec '18	NA	2d	6d

15. You should notice that the **Status Check** column changes as you updated the **Create Technical Specification** task.

16. Your updated tasks should look like this:

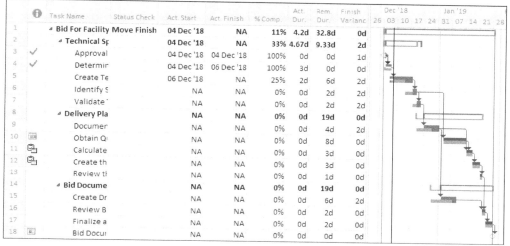

	Task Name	Status Check	Act. Start	Act. Finish	% Comp.	Act. Dur.	Rem. Dur.	03 Dec '18 S M T W T F S S
1	◢ Bid For Facility Move Finish		04 Dec '18	NA	11%	4.2d	32.8d	
2	◢ Technical Sp		04 Dec '18	NA	33%	4.67d	9.33d	
3	✓ Approval		04 Dec '18	04 Dec '18	100%	0d	0d	
4	✓ Determir		04 Dec '18	06 Dec '18	100%	3d	0d	
5	Create Te		06 Dec '18	NA	25%	2d	6d	

17. Change the timescale back to the previous settings of month and week.

18. Your schedule should look like the picture below:

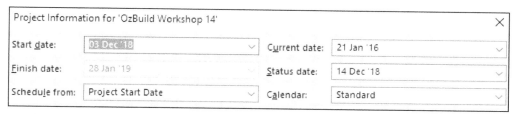

	Task Name	Status Check	Act. Start	Act. Finish	% Comp.	Act. Dur.	Rem. Dur.	Finish Varianc	Dec '18 26 03 10 17 24	Jan '19 31 07 14 21 28
1	◢ Bid For Facility Move Finish		04 Dec '18	NA	11%	4.2d	32.8d	0d		
2	◢ Technical Sp		04 Dec '18	NA	33%	4.67d	9.33d	2d		
3	✓ Approval		04 Dec '18	04 Dec '18	100%	0d	0d	1d		
4	✓ Determir		04 Dec '18	06 Dec '18	100%	3d	0d	0d		
5	Create Te		06 Dec '18	NA	25%	2d	6d	2d		
6	Identify S		NA	NA	0%	0d	2d	2d		
7	Validate		NA	NA	0%	0d	2d	2d		
8	◢ Delivery Pla		NA	NA	0%	0d	19d	0d		
9	Documer		NA	NA	0%	0d	4d	2d		
10	Obtain Q		NA	NA	0%	0d	8d	0d		
11	Calculate		NA	NA	0%	0d	3d	0d		
12	Create th		NA	NA	0%	0d	3d	0d		
13	Review th		NA	NA	0%	0d	1d	0d		
14	◢ Bid Docume		NA	NA	0%	0d	19d	0d		
15	Create Dr		NA	NA	0%	0d	6d	2d		
16	Review B		NA	NA	0%	0d	2d	0d		
17	Finalize a		NA	NA	0%	0d	2d	0d		
18	Bid Docur		NA	NA	0%	0d	0d	0d		

19. **NOTE:** there is no change in the end date of the project as there is sufficient Float to absorb the delay. We will now update the schedule without using the **Update Progress** function.

20. Move the **Status Date** to 14 Dec 18 by selecting **Project**, **Properties** group, **Project Information**.

Project Information for 'OzBuild Workshop 14'			✕
Start date:	03 Dec '18	Current date:	21 Jan '16
Finish date:	28 Jan '19	Status date:	14 Dec '18
Schedule from:	Project Start Date	Calendar:	Standard

21. Notice the **Status Check** column for task 5 indicates an error and makes a suggestion on how to fix this issue.

22. Update the project detail tasks only with the following information. Enter the data in the exact order below against the detail tasks only. **NOTE:** Do not edit the % Comp, the software will calculate this:

   ➤   **Completed Tasks** enter the Actual Start and Actual Finish for, e.g. Task 5.

   ➤   **In-progress Tasks** enter the Actual Start, Actual Duration, so the Actual Duration meets the **Status Date** and then the Remaining Duration, Tasks 6 and 7.

   ➤   The % Complete of in-progress tasks will calculate automatically.

		Task Name	Act. Start	Act. Finish	Act. Dur.	Rem. Dur.
5	✓	Create Technical Specification	06 Dec '18	14 Dec '18	7d	0d
6		Identify Supplier Components	13 Dec '18	NA	2d	4d
7		Validate Technical Specification	14 Dec '18	NA	1d	11d

23. Add the **Total Float** (Total Slack) column.

24. Your project should look like the following pictures, without a delay to the end of the project as there is sufficient Float to absorb the delay:

		Task Name	Status Check	Act. Start	Act. Finish	% Comp.	Act. Dur.	Rem. Dur.	Finish Variance	Total Slack
1		⊿ **Bid For Facility Extension**	Move Finish	04 Dec '18	NA	23%	8.44d	28.56d	0d	0d
2		⊿ **Technical Specification**		04 Dec '18	NA	46%	9.29d	10.71d	8d	3d
3	✓	Approval to Bid		04 Dec '18	04 Dec '18	100%	0d	0d	1d	0d
4	✓	Determine Installatio		04 Dec '18	06 Dec '18	100%	3d	0d	0d	0d
5	✓	Create Technical Spec		06 Dec '18	14 Dec '18	100%	7d	0d	1d	0d
6		Identify Supplier Com		13 Dec '18	NA	33%	2d	4d	3d	-4d
7		Validate Technical Sp		14 Dec '18	NA	8%	1d	11d	8d	3d
8		⊿ **Delivery Plan**		NA	NA	0%	0d	14d	0d	0d
9		Document Delivery M		NA	NA	0%	0d	4d	8d	3d
10	▦	Obtain Quotes from S		NA	NA	0%	0d	8d	0d	0d
11		Calculate the Bid Estir		NA	NA	0%	0d	3d	0d	0d
12		Create the Project Sch		NA	NA	0%	0d	3d	0d	0d
13		Review the Delivery F		NA	NA	0%	0d	1d	0d	0d
14		⊿ **Bid Document**		NA	NA	0%	0d	13d	0d	0d
15		Create Draft of Bid Do		NA	NA	0%	0d	6d	8d	3d
16		Review Bid Document		NA	NA	0%	0d	2d	0d	0d
17		Finalize and Submit B		NA	NA	0%	0d	2d	0d	0d
18	▦	Bid Document Submit		NA	NA	0%	0d	0d	0d	0d

25. Task 6 has negative Total Float because Task 7 (its successor) has started before Task 6 has finished. If you apply the **Gantt Chart Inc Total and Neg Float** View and then you will see the **Negative Float** bar:

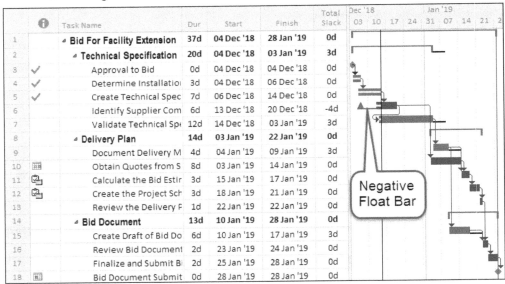

		Task Name	Dur	Start	Finish	Total Slack
1		⊿ **Bid For Facility Extension**	37d	04 Dec '18	28 Jan '19	0d
2		⊿ **Technical Specification**	20d	04 Dec '18	03 Jan '19	3d
3	✓	Approval to Bid	0d	04 Dec '18	04 Dec '18	0d
4	✓	Determine Installatio	3d	04 Dec '18	06 Dec '18	0d
5	✓	Create Technical Spec	7d	06 Dec '18	14 Dec '18	0d
6		Identify Supplier Com	6d	13 Dec '18	20 Dec '18	-4d
7		Validate Technical Sp	12d	14 Dec '18	03 Jan '19	3d
8		⊿ **Delivery Plan**	14d	03 Jan '19	22 Jan '19	0d
9		Document Delivery M	4d	04 Jan '19	09 Jan '19	3d
10	▦	Obtain Quotes from S	8d	03 Jan '19	14 Jan '19	0d
11		Calculate the Bid Estir	3d	15 Jan '19	17 Jan '19	0d
12		Create the Project Sch	3d	18 Jan '19	21 Jan '19	0d
13		Review the Delivery F	1d	22 Jan '19	22 Jan '19	0d
14		⊿ **Bid Document**	13d	10 Jan '19	28 Jan '19	0d
15		Create Draft of Bid Do	6d	10 Jan '19	17 Jan '19	3d
16		Review Bid Document	2d	23 Jan '19	24 Jan '19	0d
17		Finalize and Submit B	2d	25 Jan '19	28 Jan '19	0d
18	▦	Bid Document Submit	0d	28 Jan '19	28 Jan '19	0d

26. Increase the timescale to daily and set the **Size** in the **Timescale** form to 70%.

27. Select **File**, **Options**, **Schedule**, **Scheduling options for this project:** and check the **Split in-progress tasks** box:

28. Task 7 has **Split** as the Remaining Duration is acknowledging the predecessor and Non Work time over the Christmas holidays.

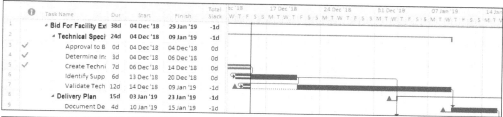

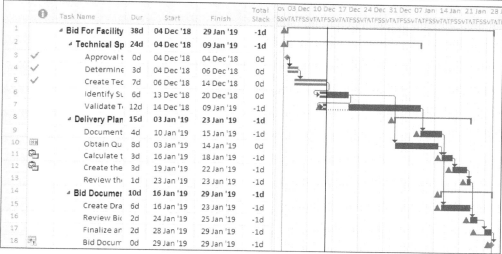

29. The project is now one day late and represented by the Negative Float of one day and the Finish Variance of one day.

30. What options do you have to bring the project back on schedule? Use your option to remove the negative float.

31. Save your **OzBuild Bid** project.

17   CREATING RESOURCES AND COSTS

A resource may be defined as something or someone that is assigned to a task and is required to complete the task. This includes people or groups of people, materials, equipment and money. The following reasons may be driving the use of resources in a schedule and each will command a different approach to the creation and updating of a resourced schedule:

- Estimating the cost or resource requirements of a project,

- Creating a platform for recording effort through time sheeting,

- Forecasting future resource requirements, with and without effort recording and

- For cash flow forecasting and/or Earned Value Performance Management.

When it is planned to update a resourced schedule it is recommended that the minimum number of resources be assigned to tasks. Avoid cluttering the schedule with resources that are in plentiful supply or are of little importance. Every resource added to the schedule will need to be updated. Therefore the scheduler's workload increases as resources are added to tasks.

When you create your resources, you should consider them under the following headings:

- Resources used as **Input Resources** – Those that are required to complete the work:
  - ➢ Individual people by name.
  - ➢ Groups of people by trade or skill.
  - ➢ Individual equipment or machinery by name.
  - ➢ Groups of resources such as Crews or Teams made up of equipment and machinery.
  - ➢ Materials or Money.

- Resources used as **Output Resources** – Things that are being delivered or produced:
  - ➢ Specifications completed.
  - ➢ Bricks laid.
  - ➢ Lines of code written.
  - ➢ Tests completed.

After planning your resource requirements, the following steps should be followed to create and use resources in a Microsoft Project schedule:

- Create your resources in the **Resource Sheet**.
- Manipulate the resource calendar if resources have special timing requirements.
- Assign the resources to tasks.
- Review the resource requirements through **Resource Graph** and **Resource Usage** views.

This chapter will concentrate on:

- The creation of resources in the **Resource Sheet**,
- Understanding **Task Type** and **Effort-Driven Tasks**, and
- Editing **Resource Calendars**.

Resources may be shared between projects using the **Resource**, **Assignments** group, **Resource Pool**, **Share Resources…** function or using the Project Server Enterprise Resource pool. These are both out of the scope of this book.

17.1 Creating Resources in the Resource Sheet

To add resources to the Resource Sheet, select **View**, **Resource Views** group, **Resource Sheet**.

	❶	Resource Name	Type	Material Label	Initials	Group	Max. Units	Std. Rate	Base Calendar	Accrue At	Ovt. Rate	Cost/Use
1		Project Manager	Work		PM	Office	100%	$120.00/hr	Standard	Prorated	$0.00/hr	$0.00
2		Systems Engineer	Work		SE	Office	100%	$90.00/hr	Standard	Prorated	$0.00/hr	$0.00
3		Project Support	Work		PS	Site	100%	$80.00/hr	6-Day Working Week	Prorated	$0.00/hr	$0.00
4		Purchasing Officer	Work		PO	Office	100%	$70.00/hr	Standard	Prorated	$0.00/hr	$0.00
5		Clerical Support	Work		CS	Office	100%	$50.00/hr	Standard	Prorated	$0.00/hr	$0.00
6		Specialist Consultant	Cost		SC	Contractor				Prorated		
7		Report Binding	Material	each	RB			$100.00		Prorated		$0.00

- To add a new resource:
  - ➢ Click on the first blank line and enter the resource data, or
  - ➢ Highlight the line where you would like to insert your resource, right-click and select **Insert Resource**, or
  - ➢ Use the **Ins Key**.

- The first column, with a ⓘ in the header, is the indicator column; 🗒 is the Notes indicator; and ◈ is the resource over-allocation indicator.

- Enter the **Resource Name**. Microsoft Project allows duplicate Resource Names to be entered into the Resource Table, but this situation should be avoided by visually checking all your Resource Names. When resources are to be shared amongst projects the spelling and coding of the resources must be carefully planned.

- Enter one of the three Resource **Types**: **Work** or **Materials** or **Cost**. They function differently:
  - ➢ **Work** resources, such as people, often have a limit to the number that are available. This type of resource may have a maximum number assigned, a standard and an overtime rate and may also have a calendar assigned.
  - ➢ **Materials** resources have a **Material Label** which is not available with a **Work** resource, have a quantity but do not have a maximum availability. They are considered unlimited and do not have a calendar.
  - ➢ **Cost** resources have no quantity or calendar but may be graphed.

- A **Material** resource may be assigned a **Material Label**, which may be the unit of measurement, such as m³ or feet.

- The **Initials** are filled out by Microsoft Project with the first character of the **Resource Name**. This Initial field should be unique. Because it is displayed in some Views. It should be edited so it makes sense to you. For example, you could use a person's initials, log-in name or payroll number.

- You may enter any text in the **Group** field if you want to group, filter or sort your resources by this field. This could be used for Department, Branch or Skill.

- The method of entering the **Resource Units** may be formatted as a percentage or as a decimal via options. Enter the **Max Units** for **Work** resources only. 100% (or 1.00) would represent one person and 400% (or 4.00) would represent four people. This value is used in Usage views to display overloading.

- Select the **File**, **Options**, **Schedule** tab to set your option of Percentage or Decimal, 2 or 200% is two resources:

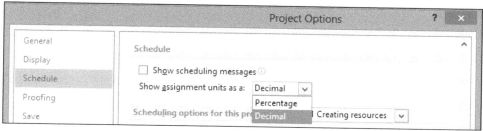

- Edit the **Standard Rate** for **Work** and **Material** resources only, if required. The resource default units are "hours" (hr), but this may be edited to be the cost per:

  ➤ minute – min

  ➤ hour – hr

  ➤ day – day

  ➤ week – wk

  ➤ month – mon

- Enter the **Overtime Rate** for the **Work** resources only. A salaried person may not be paid overtime and would have a zero rate for overtime.

- Enter the **Cost per Use** for **Work** and **Material** resources only. This could represent a mobilization cost and is applied each time a resource is assigned to a task. The **Cost per Use** is accrued when an actual start date is entered.

- **Accrue At**:

  ➤ **Start** – The costs for this resource are incurred when the task commences.

  ➤ **Prorated** – The costs for this resource are spread over the duration of the task.

  ➤ **End** – The costs for this resource are incurred when the task is complete.

- **Base Calendar** – When a new resource is created it is assigned the **Project Base** Calendar and may be changed. This is the calendar assigned to **Work** resources and may be edited to suit each individual resource availability including editing for individual holidays. See paragraph 17.4.2

17.2   Grouping Resources in the Resource Sheet

Resources may be grouped on any data fields, such as Custom Fields and Custom Outline Codes, using the **View**, **Data**, **Group by:** function. The example below shows resources grouped by **Resource Group**:

	Resource Name	Type	Materia Label	Initials	Group	Max. Units	Std. Rate	Base Calendar	Cost/Use	Ovt. Rate	Accrue At
	◢ **Type: Work**	**Work**				5			$0.00		
1	Project Manager	Work		PM	Office	1	$120.00/hr	Standard	$0.00	$0.00/hr	Prorated
2	Systems Engineer	Work		SE	Office	1	$90.00/hr	Standard	$0.00	$0.00/hr	Prorated
3	Project Support	Work		PS	Site	1	$80.00/hr	Standard	$0.00	$0.00/hr	Prorated
4	Purchasing Officer	Work		PO	Office	1	$70.00/hr	Standard	$0.00	$0.00/hr	Prorated
5	Clerical Support	Work		CS	Office	1	$50.00/hr	Standard	$0.00	$0.00/hr	Prorated
	◢ **Type: Material**	**Material**							$0.00		Prorated
7	Report Binding	Material	each	RB			$100.00		$0.00		Prorated
	◢ **Type: Cost**	**Cost**									Prorated
6	Specialist Consultant	Cost		SC	Contractor						Prorated

17.3 Resource Information Form

The **Resource Information** form is opened by double-clicking on a specific row within the **Resource Sheet** view.

17.3.1 General Tab

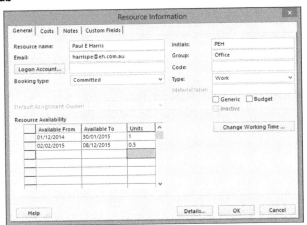

- The **Resource name:** is the name of the resource and may be displayed in columns with assignment **Units** (per time period). It is best to keep this name as short as possible to enable it to be easily displayed in columns.

- The **Email** field is used in email communication and is not covered in this book.

- [Logon Account...] allows the assignment of a resource to a Windows account and is not covered in this book.

- **Booking Type:** is an option to tag the assignment as **Committed** or **Proposed**. A **Proposed** resource does not affect the availability of a resource.

- **Resource Availability** is used to identify the resource availability over time and multiple periods may be created with different availability in each period.

- **Initials:** should be kept unique to make filters run properly and prevent corrupt files. This field may be displayed in columns in the Gantt chart, but when displayed in columns it will not display the resource **Units** but the **Resource Name** column will display the Units, therefore this column is not as useful as the **Resource Name** field in the Gantt Chart view.

- **Group** and **Codes** are fields that may be used to group resources by any user defined description such as Office, Site, City, etc.

- **Work** has three options discussed earlier, **Work**, **Material** or **Cost**.

- **Generic** resources are used to indicate long term resource requirements and may be substituted with people as the work appears on the immediate horizon.

- **Budget** indicates that the resource is being used for budget purposes and the **Resource Availability** features are disabled when a resource is marked as Budget. New to Microsoft Project 2007.

- **Inactive** in an enterprise environment indicates the resource is no longer available but may be assigned to tasks in the past so it is preferable not to delete the resource. They are excluded from assignment lists.

- The [Change Working Time ...] button opens the **Change Working Time (Resource)** form editing the resource calendar. Select each resource's **Base calendar** and edit the calendar to suit the resource's individual availability dates and times.

17.3.2 Costs Tab

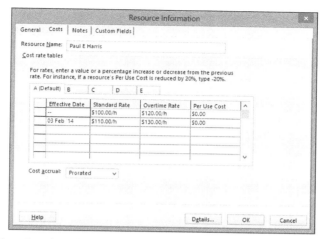

The **Cost rate tables** allow the adjustment of resource rates over time and allows the assignment of up to five different rates to a resource. The rate is assigned in the **Resource Information** form. This form is opened by double-clicking on the **Resource Name** for any assigned resource in either the **Resource Sheet**, **Task Usage** view or the **Resource Usage** view.

17.3.3 Notes

This is where notes are made about the resource such as assumptions.

17.3.4 Custom Fields

Resource Custom Fields are for assigning user defined information to resources and are not covered in detail in this book. This could include personal information such as address, qualifications, office location etc. These are different to the Task Custom Fields.

- **Custom Resource** fields may be created in the **Custom Fields** form by selecting **Project**, **Properties** group, **Custom Fields**.

- After the fields have been defined in the **Custom Fields** form, these fields appear in the **Resource Information** form and may be used to assign custom-defined costs and other information to resources from the **Custom Fields** tab of the **Resource Information** form or by adding columns to views such as the **Resource Sheet**.

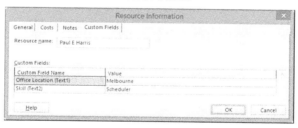

17.4   Resource Calendars

 This whole subject is extremely complex once users start using task and/or resource calendars, users must be careful that they understand the relationship amongst the calendars once they start using task and/or resource calendars with different work days and/or hour. I strongly recommend that a single calendar should be considered for all task and resources and then the software becomes more predictable.

Tasks calendars may not accommodate specific resource availability. For example, a Project Calendar would not usually reflect when a person takes a holiday. Resource Calendars, on the other hand, can be used to schedule this resource-specific nonworking time. Therefore Resource Calendars could be used to schedule work when specific resources have a unique availability.

A unique resource calendar is created automatically when a resource is created and is a copy of the project **Base Calendar** when the resource is created.

When a resource is assigned to a task in the situation where the Project Calendar, Task Calendar and Resource Calendar are all different; then the work is calculated based on some very complex parameters. This topic is extremely complex and is covered in more detail below.

17.4.1   Editing a Resource calendar

The Resource Calendar may be edited in the same manner as Base Calendars from the **Change Working Time** form. This form may be opened by opening the **Resource Information** form (double-clicking on a **Resource Name** in most forms) and selecting the [Change Working Time ...] in the **Resource Information** form, **General** tab.

17.4.2   Resource Calendars Calculations

The hierarchy of Calendars is interesting and is documented below. The table below summarizes this hierarchy.

NO RESOURCES ASSIGNED TO TASKS

- The **Project Calendar** is used to calculate the Task Dates and Task Duration when no Task Calendar or resources are assigned.

- **Task Calendar** is used to calculate the Task Dates and Task Duration when a Task Calendar is assigned and no resources are assigned.

RESOURCES ASSIGNED AND NO TASK CALENDAR ASSIGNED

- The **Resource Calendar** is used to schedule resource work when resources are assigned to an activity without a **Task Calendar**.

  - ➤ **Fixed Units** and **Fixed Work** tasks calculate the Task Dates, Task Duration using the **Resource Calendar** only. The **Project Calendar** is not considered.

  - ➤ **Fixed Duration** tasks calculate the Task Dates and Task Duration using the **Project Calendar** and the resources work when their **Resource Calendar** allows, within the Task Duration.

A **Fixed Duration** task will schedule work when the task duration spans over a period that the resource may work. Work will be scheduled when the **Task Calendar** and **Project Calendar** have common working time as long as the Task spans a Resource Work availability period. Thus, if a **Project Calendar** has Monday to Friday working and the **Resource Calendar** has a Saturday and Sunday working, then the work may not be scheduled and you will receive an error message. But if a **Project Calendar** has Monday to Friday working and the **Resource Calendar** has Friday to Sunday as working then the work will be scheduled only on Friday.

RESOURCES ASSIGNED AND TASK CALENDAR ASSIGNED

- Resource Work is scheduled using a combination of the remaining available work time from the **Resource Calendar** and **Task Calendar**. For example a resource that is only available Friday and Saturday and assigned to a Task with a Monday to Friday calendar will only have Work scheduled on Fridays.

  > **Fixed Units** and **Fixed Work** tasks calculate Task Dates and Task Duration based on the **Resource Calendars** only. The **Task Calendar** is ignored for date calculation and is only used to allow the scheduling of work in combination with the Resource Calendar.

  > **Fixed Duration** calculates the Task Dates and Task Duration using the **Task Calendar**.

 A task may not be scheduled and an error message will be displayed when a task is **Fixed Duration** and there is no common available working time from the **Resource Calendar** and **Task Calendar**. This would happen when a task is assigned a **Task Calendar** with Monday to Friday working and the **Resource Calendar** working on Saturday and Sunday.

- **Scheduling ignores resource calendars** – when checked the **Resource Calendar** is ignored and **Task Calendar** is used to calculate the Task Dates, Task Duration and schedule Work. **NOTE:** This option is not available when a Task Calendar has not been assigned.

Task Calendar	Resource assigned	Scheduling ignores resource calendars	Task Type	Calendar Used to Calculate Work	Calendar Used to Calculate Dates
None	No	N/A	Any	N/A	Project
Yes	No	N/A	Any	N/A	Task
None	Yes	N/A	Fixed Units or Fixed Work	Resource	Resource
None	Yes	N/A	Fixed Duration	Resource	Project
Yes	Yes	No	Fixed Units or Fixed Work	Remaining Availability of the combination of the Task and Resource.	Resource
Yes	Yes	No	Fixed Duration	Remaining Availability of the combination of the Task and Resource.	Task
Yes	Yes	Yes	Any	Task	Task

17.4.3   Issues Using Resource Calendars

The user must be aware that:

- The assignment of a Resource with a calendar that is different to the Task Calendar will change the task dates, durations and float values.

- The option to ignore Resource Calendars when no Task Calendar has been assigned is not available. So when a Task Calendar is NOT assigned, the Resource Calendar will always be acknowledged and the schedule dates will often change when assigning a resource with a different calendar to the Project Calendar. To simplify the schedule calculation you may consider ensuring all Resource Calendars are the same as the Task Calendars.

- The Start and Finish dates of a Fixed Duration task may be different to the Start and Finish of the scheduled work. This is often longer when the Resource Calendars are different to the Task Calendars. The Start and Finish Dates therefore represent when the work may be completed and resources are scheduled within the task duration according to their calendar availability. Durations may be longer and Float shorter than Fixed Units or Fixed Work tasks.

- The Start and Finish dates of a Fixed Units or Fixed Work task are the same as the Start and Finish of the work. If you wish to display bars that depict when the work is to start and finish, then this option may be appropriate.

- The difference between the calculation of Fixed Duration, Fixed Units and Fixed Work tasks is significant and very complex; changing these options may have a significant effect of schedule calculations. Especially when changing Resource assignments.

- Increasing or decreasing the duration of tasks or changing the assignment of resources will calculate different values depending of the Task Type and the task Effort Driven setting.

- A split may be created at the end of a task when changing assignments or Task Calendars. This is represented by dots at the end of the task and may be removed by dragging the end of the task.

- The order that data is entered may result in different work or duration values; which in turn leads to confusion of inexperienced users. Try to be consistent in the order that data is entered.

- Users must spend considerable time playing with the software in order to master these complex options. You may wish to consider creating a schedule with a resource that is available Saturday and Sunday and a Task calendar that is available Friday and Saturday. Then create a number of tasks which have been assigned the resource and assign each task with a different permutation of Task Types and Task Calendars of None, Standard (5days/week) and the Friday and Saturday calendar. Display the Task Usage form to see how the resource Work is scheduled.

- Users must be careful when creating and assigning resources. To simplify the scheduling process you may wish to only use one or two set of parameters on all your tasks that you understand and not try to master them all.

- The author recommends the following set of options should be considered for basic schedules when assigning Resources:

  ➢ Fixed Units, Non Effort Driven Tasks,

  ➢ Not to use Task Calendars, except by exception,

  ➢ Ensuring all Resource Calendars are the same as the Project Calendar.

18 Workshop 15 - Defining Resources

Background

The resources must now be added to this schedule. Since we have updated our project, we need to revert to the original schedule that we saved prior to updating the current schedule.

Assignment

1. Save and close your current project.

2. Open the **OzBuild With Resources**.

3. Select **File**, **Options** from the menu and from the **Schedule** tab, set **Show assignment unit as a: Decimal**. Click "OK"

4. Open the **Resource Sheet** view, by right-clicking in the band on the left-hand side of the screen, and add the resources displayed in the picture below to the project, leaving **Cost/Use** and **Ovt.Rate** (Overtime Rate) as zero and **Accrue At** as Prorated.

5. The Project Support resource will be assigned to activities that are on a 6-Day Working Week calendar, so these resources should be assigned a 6-Day Working Week calendar so the schedule calculates the same dates.

6. **NOTE:** The $ has been displayed as currency in these workshops, so you should use your default currency.

	Resource Name	Type	Mater Label	Initials	Group	Max. Units	Std. Rate	Accrue At	Base Calendar
1	Project Manager	Work		PM	Office	1	$120.00/h	Prorated	Standard
2	Systems Engineer	Work		SE	Office	1	$90.00/h	Prorated	Standard
3	Project Support	Work		PS	Site	1	$80.00/h	Prorated	6 Day Week
4	Purchasing Officer	Work		PO	Office	1	$70.00/h	Prorated	Standard
5	Clerical Support	Work		CS	Office	1	$50.00/h	Prorated	Standard
6	Specialist Consultant	Cost		SC	Contractor			Prorated	
7	Report Binding	Material	Each	RB	Material		$100.00	Prorated	

7. Now select **View**, **Data** group, **Group by: Resource Group** to see the resource grouping function:

	Resource Name	Type	Mater Label	Initials	Group	Max. Units	Std. Rate	Accrue At	Base Calendar
	⊿ **Group: Contractor**				**Contractor**			**Prorated**	
6	Specialist Consultant	Cost		SC	Contractor			Prorated	
	⊿ **Group: Material**				**Material**			**Prorated**	
7	Report Binding	Material	Each	RB	Material		$100.00	Prorated	
	⊿ **Group: Office**				**Office**	**4**			
1	Project Manager	Work		PM	Office	1	$120.00/h	Prorated	Standard
2	Systems Engineer	Work		SE	Office	1	$90.00/h	Prorated	Standard
4	Purchasing Officer	Work		PO	Office	1	$70.00/h	Prorated	Standard
5	Clerical Support	Work		CS	Office	1	$50.00/h	Prorated	Standard
	⊿ **Group: Site**				**Site**	**1**		**Prorated**	
3	Project Support	Work		PS	Site	1	$80.00/h	Prorated	6 Day Week

8. Open the **Gantt Chart** view.

9. Save your **OzBuild With Resources** project.

---

19 ASSIGNING RESOURCES AND COSTS TO TASKS

Microsoft Project has the following types of costs and/or resources:

- **Fixed Costs** are costs allocated to tasks (without assigning a resource) in order to calculate the total cost of a project. A Cost Resource has more options and will give a better result.

- **Work** may be assigned to a task without a resource in order to calculate the total number of hours required for a project without adding a resource and display in the **Task Usage** form.

- **Work resources** are used for people or equipment. The **Work** (quantity) is usually measured in hours and the **Work** resource components are linked to the Task Duration. These resources may be assigned a **Rate** to calculate their cost and a calendar.

- **Material resources**, which have a Quantity and a Unit Rate to calculate a costs. The quantity is not linked to the task duration and may not be assigned a calendar.

- **Cost resources** were introduced in Microsoft Project 2007. This is a resource that does not have a quantity, is not affected by the task duration and may not be assigned a calendar.

Resources may be assigned to tasks using a number of methods including:

Method	Menu Command
• **Fixed Costs** assignment	Display the **Fixed Costs** and **Fixed Costs Accrual** columns.
• **Work** assignment	To assign work to a task but not assign a resource, you may simply display the **Work** column and type in the hours. A resource assigned to a task with work will inherit the work value assigned to the task.
• **Resource Assignment** form	You may assign **Units** and **Costs** by clicking on the **Resource, Assignments Group**, **Assign Resources** button ⬚.
• **Task Information** form	Double-click on a **Task** name or click on the **Task Information** button ⬚. You may assign **Resource Units** and **Costs** from the resources tab.
• The **Resource Task Details** forms	In the bottom window, select the **Task Details Form**, **Task Form** or **Task Name Form**, and then select the appropriate form from the **Details** option.
• The **Resources Names** and **Resource Initials** columns	Insert appropriate columns by right-clicking on the column heading where you want to insert a column and then assign a Resource.
• **Contoured** Resource assignment	Open the **Assignment Information** form by: • Double-clicking on a resource in the **Task Usage** or **Resource Usage** view, or • Right-clicking on a resource and selecting **Assignment Information** from the menu.
• **Shared Resources**	Select **Resource, Assignments** group, **Resource Pool, Share Resources...** to open the **Share Resources** form to share resources with other projects.

 Users may have to re-apply a view to allow some column data to recalculate.

19.1 Fixed Costs

Fixed costs are a function whereby you may assign costs to a task without creating resources. It is a useful function if you require a cash flow only but not so useful with a progressed schedule as the Actual and Remaining costs of a **Prorated Fixed Cost** are linked and proportional to the Actual and Remaining duration.

- A fixed cost is assigned using the **Fixed Cost** column.

- The fixed cost may be accrued at the **Start**, **End** or **Prorated** over the duration of the task. This option is selected from the **Fixed Cost Accrual** column. The default for the **Fixed Cost Accrual** is set in the **File**, **Options, Schedule** tab:

Task Name	Duration	Fixed Cost	Fixed Cost Accrual	M	T	W	T (03 February)	F
Summary	5 days	$0.00	Prorated	$240.00	$40.00	$40.00	$40.00	$240.00
$200.00 Fixed cost accrued at Start	5 days	$200.00	Start	$200.00				
$200.00 Fixed cost Prorated	5 days	$200.00	Prorated	$40.00	$40.00	$40.00	$40.00	$40.00
$200.00 Fixed cost accrued at End	5 days	$200.00	End					$200.00

- Fixed Costs will not be displayed in the **Resource Sheet** or **Resource Usage** views but are available in the **Task Usage** view and in a **Report**. The **Cash Flow** report will only give you the cash flow in weeks but may be exported to Excel as a Pivot Table.

- Fixed Costs are added to resource costs and the total is shown in the **Cost** column.

Task Name	Fixed Cost	Fixed Cost Accrual	Resource Initials	Cost
◢ **Summary**	$0.00	Prorated		$8,800.00
$200.00 Fixed cost accrued at Start	$200.00	Start		$200.00
$200.00 Fixed cost Prorated	$200.00	Prorated		$200.00
$200.00 Fixed cost accrued at End	$200.00	End		$200.00
Activity With Resources Only	$0.00	Prorated	PEH	$4,000.00
Activity With Costs & Resources	$200.00	Prorated	PEH	$4,200.00

- Normally the Actual Costs and remaining Costs are linked to the % Complete:

% Complete	Fixed Cost	Actual	Remaining	Total Cost
30%	$100.00	$30.00	$70.00	$100.00

- When the **File**, **Options, Schedule, Calculation Options for this project**: **Actual costs are always calculated by Project** is unchecked the:

  ➢ The Actual Costs may be entered by the user at any value, and

  ➢ The **Fixed Costs** no longer equal the **Total Costs**:

% Complete	Fixed Cost	Actual	Remaining	Total Cost
30%	$100.00	$200.00	$70.00	$270.00

When both **Fixed Costs** and **Resource Costs** are assigned to a task, there is no column to display only the total of either the **Fixed Cost** or the **Resource Cost** as they are added together in the costs column It is therefore recommended to be able to clearly understand the contents of a Cost column that either Fixed Costs or Resource Costs be assigned to a task but not both.

When a Baseline is set, the Fixed costs and the Resource costs are added together.

The Cost resource has more functionality and should be used in place of Fixed Costs.

19.2 Assigning Work without a Resource

To assign work to a task which does not have resources; display the **Work** column and type in the hours. A resource assigned to a task with work will inherit the work value assigned to the task.

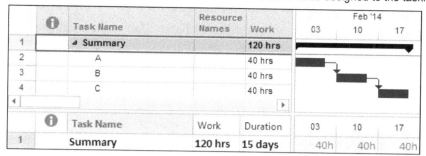

19.3 Resource Definitions

When a resource is assigned to a task, it has three principal components:

- **Quantity**, in terms of **Work** or **Material** required to complete the task,
- **Units**, which represents the number of people working on a task or material quantity and
- **Cost**, calculated from the **Standard Rate**, **Overtime Rate** and **Cost per Use**.

The **Units** (per Time Period) of a **Work** resource may be entered against a task, the **Work** (Quantity) will be calculated, or the **Work** entered and the **Units** calculated. The resource cost is calculated from the resource **Work** times the resource **Rate**.

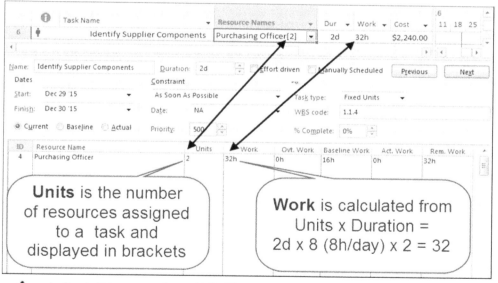

Units is the number of resources assigned to a task and displayed in brackets

Work is calculated from Units x Duration = 2d x 8 (8h/day) x 2 = 32

In Oracle Primavera software **Units/Time** is the same as Microsoft Project **Units** and the Oracle Primavera software **Units** is the same as Microsoft Project **Work**.

19.4 Task Type and Effort-Driven

This is a very difficult concept and users should consider creating a small project with one task and a couple of resources and play with different **Task Types** and **Effort Driven** options to see how the software operates.

The **Task type** and **Effort-driven** options operate after the first **Work** resource has been assigned to a task. The relationship among the following four variables are controlled by these options:

- The number of different **Resources**, i.e., painters and testers are two different resources.
- The **Task duration**.
- The resource **Units** (per Time Period), i.e., two testers would be 2 Units per Time Period.
- The amount of **Work** to be performed, i.e., the total number of hours.

The **Task type** and **Effort-driven** functions control which variable changes when one of these four parameters is changed when **Work** resources are added or removed from a task. These options may be set in a number of forms including the **Task** and the **Task Details** form. **Material** resources, **Cost** resources and **Fixed Costs** do not change values after they have been assigned.

19.4.1 Task Type – Fixed Duration, Fixed Units, Fixed Work

There is a relationship between the **Duration** of a task, the **Work** (the number of hours required to complete a task) and the **Units** (per Time Period), (the rate of doing the work or number of people working on the task). The relationship is:

Duration x Units (per Time Period) **= Work**

There are three options for the **Default task type:** which decide how this relationship operates:

- **Fixed Duration** The **Duration** stays constant when either the **Units** (per Time Period) or **Work** is changed. If you change the **Duration**, then the **Work** changes.
- **Fixed Units** The **Units** (per Time Period) stays constant when either the **Duration** or **Work** is changed. If you change the **Units** (per Time Period), then the **Duration** changes.
- **Fixed Work** The **Work** stays constant if either **Duration** or **Units** (per Time Period) is changed. Therefore your estimate will not change when you change **Duration** or **Units** (per Time Period). If you change the **Work**, then the **Duration** changes.

19.4.2 Effort-Driven

Once a resource has been assigned to a task, the task **Effort** is the combined number of hours of all resources assigned to a task. The **Effort-driven** option decides how the effort is calculated when a resource is added or when a resource is removed to a **Fixed Units** or **Fixed Duration** task. There are two options:

- **Effort-driven** When a resource is added or removed from a task, the **Task Effort** assigned to a task remains constant. Adding or removing resources leaves the total effort assigned to a task as a constant unless all resources are removed. The work is shared amongst the resources and the resources are considered equal in value.
- **Non Effort-driven** When a resource is added to or removed from a task, the **Resource Effort** or **Work** of other resources remains constant. Adding or deleting resources increases or decreases the total task effort. This is the best option in construction projects when you wish to assign resources to crews and be able to assign each resource independently to existing resource assignments.

Some resource options to consider in the **File**, **Options**, **Schedule** tab are:

- When you do not want the **Duration** to change as the resource work is edited, set the **Default task type:** as **Fixed Duration** and **Non Effort Driven**. Each resource work is independent of all other resources and the total work remains constant when the duration is changed. The **Units per Time Period** i.e. the crew size will change if the duration is changed. The Estimate to Complete will not change.

- When you do not want the **Units (**per Time Period) to change as the duration is changed then set the **Default task type:** as **Fixed Units** and **Non Effort Driven**. Therefore the crew size will NOT change if the duration is changed but the work will change. For example a task with three resources assigned will always have three resources assigned and the work will increase or decrease as the duration is increased or decreased. The Estimate to Complete will change.

The disadvantage of **Fixed Duration** tasks is that they calculate and display progress on split tasks from the start date to the finish date of the task, ignoring the split, which results in progress that sometimes looks odd.

When tasks are Effort Driven and multiple resources are to be assigned a different result is obtained if one resource is assigned at a time to the option of assigning multiple resources at a time.

19.5   *Resource Calendars*

The simplest method of scheduling is when all tasks and resources share the same calendar. This is often not desirable and Microsoft Project allows three levels of calendars, project, task and resource.

The interaction among these calendars when resources are assigned is difficult to understand and the order that resources are assigned to tasks and assignment of task calendars may result in different calculated task durations and work. It is recommended that you practice with a small schedule with one or two tasks and resources until you are confident on how the software is calculating. The calendar rules are described in detail in paragraph 17.4 Resource Calendars Calculations.

A common issue is the situation where a task is assigned a calendar, then at a later date resources are assigned without editing the resource calendars to match the task calendars and then that the task durations calculate differently.

19.6   *Assigning Resources using the Resource Assignment Form*

Highlight one or more tasks that you want to assign resources. To open the **Assign Resources** form:

- Add the **Assign Resources** button 🔳 to the **Quick Access Toolbar**, or

- Select a task and right-click and select the 🔳 **Assign Resources...**, or

- Select **Resource**, **Assignments** group, **Assign Resources**

The **Units** represent the number of people working on a task and may be displayed as **Units** shown in the picture on the left, or as a **Percentage** shown on the right. The value of 1.00 is the same value as 100%. This option may be changed in the **Schedule** tab of the **Options** form:

 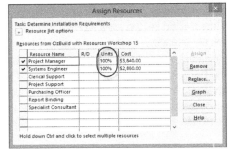

- Select the **Resource** from the form.
  - ➢ Type in the number of resources you want to assign under **Units**. This may be in % or whole numbers depending on how your Options are set.

  - ➢ Click on the [Assign] button to assign a resource to a task.

- You may then assign another resource to a task.

- Select a resource and click on the [Remove] button to remove a resource.

- Select a resource and click on the [Replace...] button to open the Replace Resource form to replace a resource.

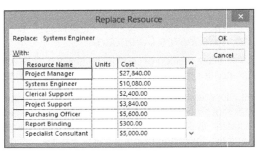

- The **Resource list options** at the top of the screen allows the resource list to be reduced by the use of filters and is very useful when there is a large number of resources or to create new resources from your address book.

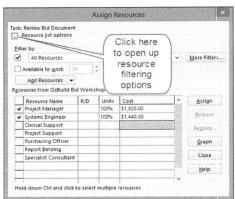

19.7 Assigning Resources Using the Task Details Form

Open the bottom window displaying the
Task Detail Form, **Task Form** or **Task Name Form**. Then select the appropriate
option by right-clicking in the form. See the
picture on the right:

The table below lists the resource
component options available in these forms:

Form	Assignment Option
• **R**esources & **P**redecessors	**Units** and **Work**
• **R**esources & **S**uccessors	**Units** and **Work**
• **S**chedule	**Work** only
• **W**ork	**Units** and **Work**
• **C**ost	**Units** only with calculated **Costs**

The **Task Form** with the Resource **Work** details form is displayed below.

- Select the required resource in the drop-down box under the **Resource Name** heading.
- Enter the number of resources under **Units**, or
- Enter the amount of **Work** – the number of hours may be entered under this heading.
- **Ovt. Work** – Overtime Work may be assigned to reduce the duration of the task. The **Work**
  remains the same, but the **Overtime** value represents the amount of work being done on
  overtime. Overtime work is not displayed as an overload in the Usage Views.

 When **Overtime** work is assigned and costs are being calculated; If an Overtime
Rate is not entered the cost will decrease as more overtime is assigned.

- **Baseline Work** is copied from **Work** when a **Baseline** is set.
- **Act. Work** – Actual Work is entered when work is in-progress.
- **Rem. Work** – Remaining Work is calculated by subtracting **Actual Work** from **Work**.

19.8 Assigning Resources Using the Task Information Form

Open the **Task Information** form:

- Select the required resource from the drop-down box.

- Type in the number of required resources in the Units box.

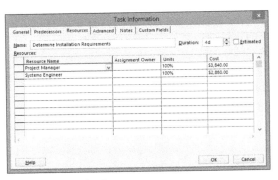

 The costs may not calculate unless you close the form.

19.9 Assigning Resources from the Resource Column

Resources may be assigned by displaying the **Resource Names** column and clicking in the cell to display the available resources. This feature was new to Microsoft Project 2010.

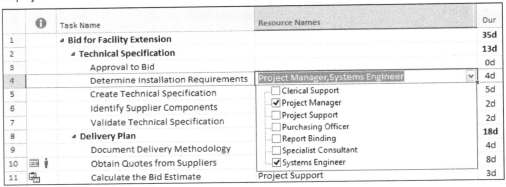

19.10 Assignment of Resources to Summary Tasks

Summary tasks may be assigned **Fixed Costs**, **Work Resources**, **Costs Resources** and **Material Resources**.

You must also be aware that when a Work resource is assigned to a summary task the task type is set to **Fixed Duration** and that setting may not be changed. Thus, any change in duration of a summary task due to rescheduling of associated detail tasks will result in a change to the work assignment and the calculated costs of a summary task.

 It is recommended that unless a summary task Work resource assignment is required to vary in proportion to the summary task duration, then Work Resources should not be assigned to a summary task. You should consider using Fixed Costs, Cost resource or a Material resource if appropriate.

19.11 Resource Engagement

Resource Engagements is a new function in Microsoft Project Professional 2016 or Project Pro for Office 365 and must be, connected to Project Online.

- If there is no connection then the Resource Engagement commands are hidden.

- A Resource Engagement allows a person to request a resource for a specific date range and percentage assignment or number of hours.

- The resource manager may then approve or suggest modifications to the resource request.

19.12 Rollup of Costs and Hours to Summary Tasks

The summary task **Cost** and **Work** fields are calculated from the sum of the costs and work assigned to the related detail tasks and those of the summary task.

 It becomes difficult for a scheduler to check the total cost of a summary task after costs and work have been assigned to both the summary task and associated detail tasks. It is therefore recommended that you consider only assigning costs and work to detail tasks.

Summary tasks have the costs and work rolled up to give you a cost at any Outline level.

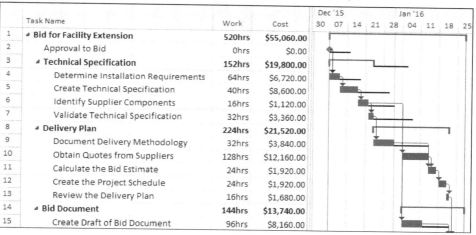

	Task Name	Work	Cost
1	⊿ **Bid for Facility Extension**	**520hrs**	**$55,060.00**
2	Approval to Bid	0hrs	$0.00
3	⊿ **Technical Specification**	**152hrs**	**$19,800.00**
4	Determine Installation Requirements	64hrs	$6,720.00
5	Create Technical Specification	40hrs	$8,600.00
6	Identify Supplier Components	16hrs	$1,120.00
7	Validate Technical Specification	32hrs	$3,360.00
8	⊿ **Delivery Plan**	**224hrs**	**$21,520.00**
9	Document Delivery Methodology	32hrs	$3,840.00
10	Obtain Quotes from Suppliers	128hrs	$12,160.00
11	Calculate the Bid Estimate	24hrs	$1,920.00
12	Create the Project Schedule	24hrs	$1,920.00
13	Review the Delivery Plan	16hrs	$1,680.00
14	⊿ **Bid Document**	**144hrs**	**$13,740.00**
15	Create Draft of Bid Document	96hrs	$8,160.00

The Baseline Costs and Work are copied from the Cost and Work fields at the time that the Baseline is set and are not calculated from their detail tasks. Thus, when tasks are added, deleted or moved to different summary tasks then the Baseline Costs and Work are no longer the sum of their detail tasks.

There is a function titled **Summary Task Interim Baseline Calculation**, which is covered in the **UPDATING PROJECTS WITH RESOURCES** chapter. This will allow the summary tasks Baseline Dates and Costs to be recalculated.

19.13 Contour the Resource Assignment

A Resource Assignment may be assigned to a task with a non-linear profile. This function is titled **Work Contour** and is similar to the Resource Curve function in Oracle Primavera software. To assign a contour to a resource assignment:

- Open the **Assignment Information** form by:
  - ➤ Double-clicking on a resource in the **Task Usage** or **Resource Usage** view, or
  - ➤ Right-clicking on a resource and selecting **Information** (Assignment) from the menu.

- The picture below shows the Sales Engineer being assigned as **Back Loaded**.

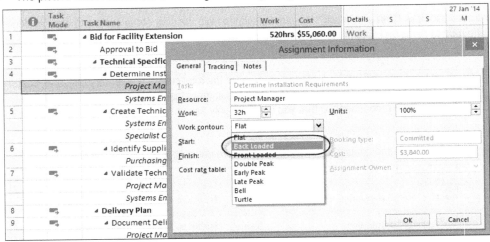

- From the **Work contour:** drop-down box, select the **Work Contour** type.

- The picture below is the **Resource Usage** form. It shows one task titled **Example Task** with nine resources. Each resource name has been made the same as the assigned **Work** contour. The picture shows:
  - ➤ The effect of the assignment for each of the **Work Contour** types, and
  - ➤ The icon in the **Task Information** column is a graphical representation of the contour.

	ⓘ	Task Name	Work	Details	M	T	W	T	F
1	ᵢ	Example Task	360 hrs	Work	47.68h	103.27h	76.1h	86.27h	46.68h
		Flat	40 hrs	Work	8h	8h	8h	8h	8h
	▮▮▮	Back Loaded	40 hrs	Work	1.67h	5h	8.33h	11.67h	13.33h
	▮▮▮	Front Loaded	40 hrs	Work	13.33h	11.67h	8.33h	5h	1.67h
	▮▮▮	Double Peak	40 hrs	Work	6h	12h	4h	12h	6h
	▮▮▮	Early Peak	40 hrs	Work	6h	16h	10h	6h	2h
	▮▮▮	Late Peak	40 hrs	Work	2h	6h	10h	16h	6h
	▮▮▮	Bell	40 hrs	Work	2.4h	9.6h	16h	9.6h	2.4h
	▮▮▮	Turtle	40 hrs	Work	4.28h	10h	11.43h	10h	4.28h
	🖉	Contoured	40 hrs	Work	4h	25h	0h	8h	3h

The **Flat** option allows you to type in any value into each of the cells. A resource assignment is automatically set to **Contoured** when one manual entry is made in the table. This is similar to **Bucket Planning** in P6.

19.14 Workshop 16 - Assigning Resources to Tasks

Background

The resources must now be assigned to their specific tasks.

Assignment

Open the **OzBuild With Resources** project and complete the following steps:

1. Select the **Gantt Chart** view.
2. Insert the **Resource Names** column to the right of the **Task Name** and align the data to the left.
3. Split the pane by right-clicking in the Gantt Chart and selecting **Show Split**.
4. Display the **Task Details Form** in the bottom pane window.
5. Right-click in the **Task Details Form** and choose **Work** from the menu.
6. **Ensure all Detail Tasks are NOT Effort Driven and are Fixed Units.**

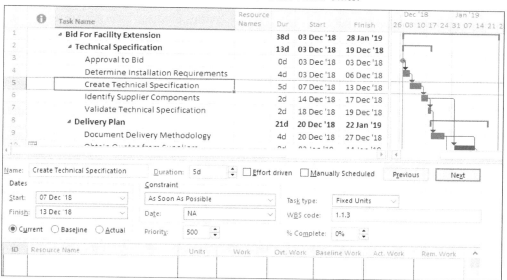

7. Assign the Resources as per table below the using any suitable method including:

   ➢ The **Task Details Form** in the bottom window or

   ➢ The **Assign Resources** form by clicking on the button 👥 or

   ➢ The drop down box in the **Resource Names** column.

NOTES: Once you have entered the **Resource Name** in the **Task Details** form and assigned the **Units**, Microsoft Project will calculate the worked hours automatically after you click out of the form or press the enter key. Ensure the tasks are **NOT Effort-Driven** as you enter resources, otherwise the Total Work may stay constant and the duration change as you add additional resources. As resources are assigned to tasks 10 and 11 the task calendars will be ignored but as the resource has been assigned a 6-Day Working Week calendar these tasks will be scheduled with the same dates.

ID	Task Name	Resources
4	Determine Installation Requirements	Project Manager, Systems Engineer
5	Create Technical Specification	Systems Engineer, Specialists Consultant and assign a cost of $5,000.00
6	Identify Supplier Components	Purchasing Officer
7	Validate Technical Specification	Project Manager, Systems Engineer
9	Document Delivery Methodology	Project Manager
10	Obtain Quotes from Suppliers	Purchasing Officer, Project Manager
11	Calculate the Bid Estimate	Project Support
12	Create the Project Schedule	Project Support
13	Review the Delivery Plan	Project Manager, Systems Engineer
15	Create Draft of Bid Document	Clerical Support, Project Manager
16	Review Bid Document	Systems Engineer, Project Manager
17	Finalize and Submit Bid Document	Project Manager, Report Binding [3 each]

8. Format your columns as per below to check your answer:

		Task Name	Resource Names	Work	Cost
1		⊿ Bid For Facility Extension		520h	$55,060.00
2		⊿ Technical Specification		152h	$19,800.00
3		Approval to Bid		0h	$0.00
4		Determine Installatior	Project Manager,Systems Engineer	64h	$6,720.00
5		Create Technical Speci	Specialist Consultant[$5,000.00],Systems Engineer	40h	$8,600.00
6		Identify Supplier Comp	Purchasing Officer	16h	$1,120.00
7		Validate Technical Spe	Project Manager,Systems Engineer	32h	$3,360.00
8		⊿ Delivery Plan		224h	$21,520.00
9		Document Delivery Me	Project Manager	32h	$3,840.00
10		Obtain Quotes from Su	Purchasing Officer,Project Manager	128h	$12,160.00
11		Calculate the Bid Estir	Project Support	24h	$1,920.00
12		Create the Project Sch	Project Support	24h	$1,920.00
13		Review the Delivery Pl	Project Manager,Systems Engineer	16h	$1,680.00
14		⊿ Bid Document		144h	$13,740.00
15		Create Draft of Bid Doc	Clerical Support,Project Manager	96h	$8,160.00
16		Review Bid Document	Systems Engineer,Project Manager	32h	$3,360.00
17		Finalize and Submit Bi	Project Manager,Report Binding[3 Each]	16h	$2,220.00
18		Bid Document Submitt		0h	$0.00

9. Now display the **Task Usage** view in the Bottom Pane. This displays the Tasks and the Resources assigned to the tasks.

10. Display the columns shown below by adding the **Cost** column,

11. Select all the tasks in the top pane and check that your data matches the table:

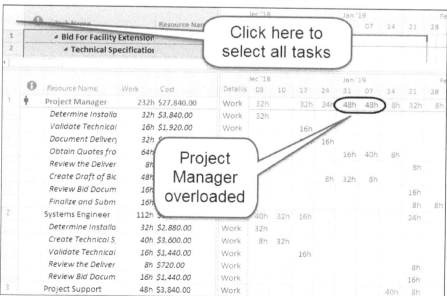

12. Now display the **Resource Usage** view in the Bottom Pane. This displays the resources and the tasks assignment for each resource.

13. Display the columns shown below by adding the Costs column,

14. Select all the tasks in the top pane and check that your data matches the table.

15. The Project Manager is overloaded where the values are in red.

16. Remove the Split and remove the resources column.

17. Save your **OzBuild Bid** project.

---

20 RESOURCE OPTIMIZATION

At this point in the preparation of the schedule the resources should be reviewed and optimized to ensure that they are being efficiently used and sufficient resources are available.

This chapter will briefly cover the following topics which will enable you to review your resources:

- **Resource Histograms** – Allow the display of each resource on a time-phased vertical bar graph. These are termed **Resource Graphs** in Microsoft Project.

- **Resource Tables** – Allow the display of one or more resource requirements in a table on a time-phased basis. This information is displayed by the **Task Usage** and **Resource Usage** views.

- **S-Curves** – Allow the display of the planned, earned or actual consumption of resources or costs as a line graph. A single S-Curve may be produced for one or selected resource costs or hours using the Resource Graph view. The display of multiple S-Curves showing the Planned, Actual and Earned cost or hours of resources is not possible. It is also not possible to display Fixed Costs on the S-Curves, but the source data for resources may be exported to a spreadsheet where multiple S-Curves may be readily created.

- **Leveling** – There are several techniques available for smoothing resource peaks or overloads in resource requirements.

20.1 Resource Graph Form

The **Resource Graph** form which was available from the **Assign Resources** form was a new feature to Microsoft Project 2002 and was removed in Microsoft Project 2010.

Selecting [Graph] from the **Assign Resources** form will now display the **Resource Graph** form in the Bottom Pane, as per the picture below and not produce a separate picture as in earlier version of Microsoft Project:

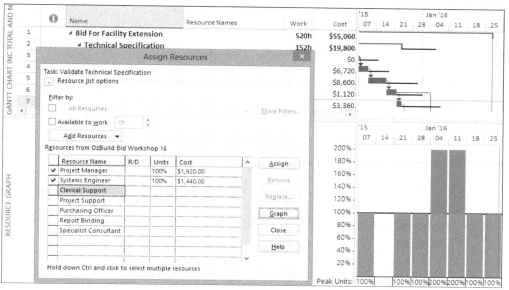

20.2   Resource Graph View

A **Resource Graph** may be displayed with the **Resource Graph** view.

The **Resource Graph** may be viewed in the top or bottom pane. The view below displays the Resource Usage in the top pane and Resource Graph in the bottom pane and shows the Project Manager is overloaded:

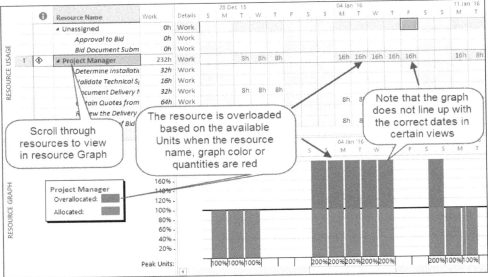

- Only one resource is displayed at a time unless specially formatted as discussed next. You may scroll through the resources by:
  - ➢ Pressing the **Page Up** and **Page Down Keys**, or
  - ➢ Clicking on the Resources in the **Resource Usage** view.

- The **Resource Graph Units** options include Cumulative Costs and Work, Costs, Work and Percentage Allocation.
- Right-clicking in the graph area of the screen to display the options.
- The **Resource Graph** gridlines and colors may be formatted by right-clicking in the graph area of the screen and selecting the **Bar Styles** option.
- The **Timescale** of the graph is adopted from the timescale setting. You will find the **Zoom** buttons useful for changing the timescale.

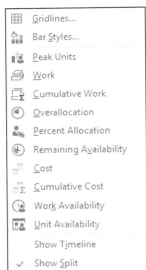

- To show a histogram for selected Work resources:

    ➢ Display the **Resource Usage** in the top pane,

    ➢ Display the **Resource Graph** in the bottom pane,

    ➢ Right-click in the **Resource Graph** and open the **Bar Styles** form,

    ➢ Ensure there is a **Bar** displayed for **Selected Resources**,

    ➢ Close the form,

    ➢ Select one or more resources in the Resource Usage view.

    ➢ The result is as per the picture below:

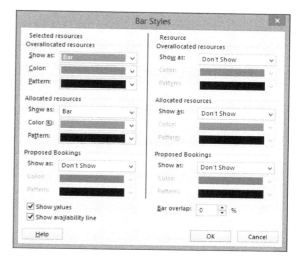

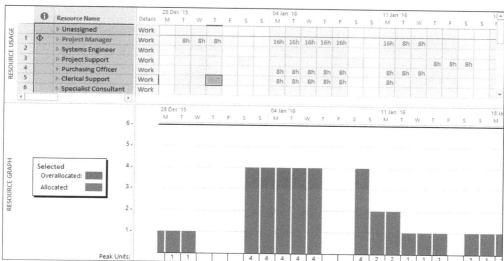

This publication is only sold as a bound book, no parts may be reproduced by any means, e.g. electronic, video or print.

© **Eastwood Harris Pty Ltd**   **253**

20.3 Resource Tables View

Resource tables are displayed using the **Task Usage** or **Resource Usage** views.

- The **Task Usage** view organizes the schedule by **Task** and then by **Resource**,

- Summarizing in the **Gantt Chart** view will summarize the **Task Usage View:**

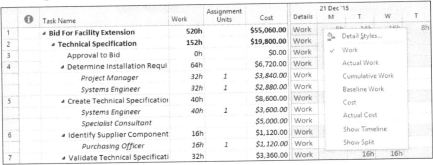

- The **Resource Usage** view organizes the schedule by **Resource** and then by **Task**. Note the Project Manager is overloaded as indicated by the icon in the indicators column:

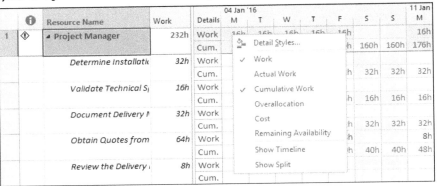

- When these views are displayed in the top **Pane** they display all resources and all tasks.

- When these views are displayed in the bottom **Pane** they display values for the resources or tasks that have been selected in the top **Pane**.

- More than one line of data may be displayed by right-clicking in the display and selecting from the menu as shown above.

- There are some interesting additional Resource Assignment fields available in this view that you may wish to experiment with.

20.4 Detailed Styles Form

This form may be opened from the **Resource Usage** and **Task Usage** forms by right-clicking and selecting 🔧 **Detail Styles...**. This allows further options for formatting these views including creating a single S-Curve:

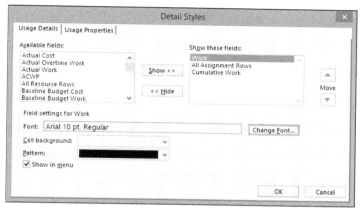

20.5 Creating an S-Curve from Microsoft Project

A single cost S-Curve for the current schedule and not the Baseline may be created graphically and displayed by Microsoft Project. The example below is produced by:

- Displaying the **Resource Sheet** in the top pane and selecting all the resources,
- Displaying the **Resource Graph** in the bottom pane,
- Right-clicking and displaying the **Cumulative Costs** in the bottom pane and
- Right-clicking, opening the 📊 **Bar Styles...** form and formatting as displayed below:

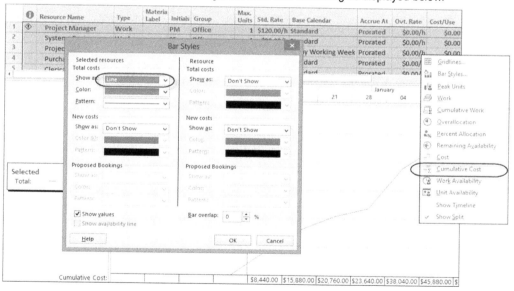

20.6 Team Planner View

The **Team Planner** view was new to Microsoft Project 2010, available in the Professional Version only and is a very useful view for analyzing a schedule. This view allows a **Production Planning** type view where tasks may be assigned to resources as opposed to **Project Planning** where resources are assigned to tasks. The picture below clearly shows where the Project Manager is overloaded as he/she is assigned to two tasks at the same time:

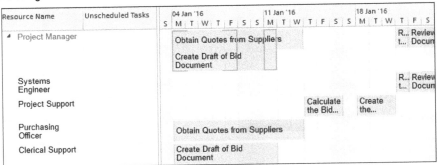

An overloaded resource may have an assignment reassigned by selecting the resource assignment in the right hand side, then right-clicking to select **Reassign To** and select an alternative resource:

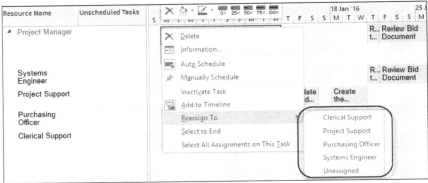

- Double-clicking on the Task will open up the **Task Information** form.

- Double-clicking on the Resource will open up the **Resource Information** form.

- Dragging a Task to another Resource will reassign the task to a resource.

- If Work is assigned to a task (without assigning a resource) using the Work column then a task will be shown as **Unassigned**:

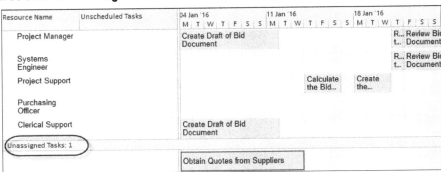

20.7   Printing Resource Profiles and Tables

To **Print** a **Task Usage**, **Resource Usage**, or **Resource Graph**, make the appropriate **Pane** active and use the print functions as described in the **PRINTING AND REPORTS** chapter.

20.8 Creating Resource Graphs, Crosstab Tables, S-Curves in a Spreadsheet

Resource Graphs, Crosstab Tables and S-Curves may be created for displaying:

- Planning information, such as the number of people required or a project cash flow,

- Progress in terms of hours spent or cost to date, and

- Performance, for example comparing planned and actual hours or costs to date.

Earned Value also known as **Earned Value Management** (**EVM** PMI term) or **Earned Value Performance Measurement** (**EVPM** AS-4817 term) is a well-documented technique used for the management of projects. This technique requires the results output in a number of formats which Microsoft Project is not able to reproduce. Spreadsheets or specialists software will produce these outputs from data exported from Microsoft Project. There are many good Earned Value reference documents that may be used to assist in the establishment of an Earned Value Performance Measurement system including:

- Australian Standard "AS 4817-2008 Project performance measurement using Earned Value", and

- Project Management Institute Practice "Standard For Earned Value Management".

Excel is often used to create S-Curves and Graphs (Histograms) for the following reasons:

- The ability to easily record and display historical period information, this may be hours consumed or dollars spent per week or per month.

- The ability to simply record and enter actuals at a summary level in Excel. It is simpler to record cost and hours at project or at WBS component than at a detail task level. Microsoft Project requires the actuals to be assigned

- The superior graphical output of Excel and flexibility compared to Microsoft Project; such as the ability to display both tabular and graphical information and multiple curves on a single printout.

Many people are familiar with exporting or copying and pasting tabular data from one software program to another and are also comfortable with creating S-Curves and Histograms in spreadsheets which are titled charts in Excel. This section of this book will cover the basics of transferring the data from Microsoft Project to a spreadsheet for the creation of S-Curves and Histograms.

20.8.1   Export Time Phased Data to Excel Using Visual Reports

In Microsoft Project 2007 the **Analysis** toolbar export function was moved to the **Visual Reports - Create** form. Reports like the **Cash Flow Report** and the **Earned Value Over Time Report** are used to export and/or display time-phased data.

You may find this function difficult to use as the data is exported in pivot table format and dates are, by default, seen as week numbers.

	A	B	C	D	E
1	Tasks	All			
2					
3				Data	
4	Year	Quarter	Week	Cost	Cumulative Cost
5	2015	Q4	Week 50	$ 8,440.00	$ 8,440.00
6			Week 51	$ 8,000.00	$ 16,440.00
7			Week 52	$ 5,440.00	$ 21,880.00
8			Week 53	$ 2,880.00	$ 24,760.00
9		Q4 Total		$ 24,760.00	$ 24,760.00
10	2015 Total			$ 24,760.00	$ 24,760.00
11	2016	Q1	Week 53	$ 1,360.00	$ 26,120.00
12			Week 1	$ 14,400.00	$ 40,520.00
13			Week 2	$ 6,480.00	$ 47,000.00
14			Week 3	$ 5,280.00	$ 52,280.00
15			Week 4	$ 3,900.00	$ 56,180.00
16		Q1 Total		$ 31,420.00	$ 56,180.00
17	2016 Total			$ 31,420.00	$ 56,180.00
18	Grand Total			$ 56,180.00	$ 56,180.00

20.8.2 Export Using Time Phased Data Copy and Paste

A **Resource Usage** view may be highlighted, copied, and then pasted into spreadsheets to create S-Curves. The Microsoft Project headings are not copied, and the data on the left-hand side of the **Pane** and the data on the right-hand side of the **Pane** have to be copied separately and aligned with manually created headings in Excel.

To transfer the data into Excel by copy and paste:

- Organize the **Task Usage** or **Resource Usage** view and **Details** so it displays the rows, columns, units and timescale you require in your graph.

| | Task Name | Work | Details | 07 Dec '15 | | | | | | | 14 Dec '15 | |
				M	T	W	T	F	S	S	M	T
1	◢ Bid For Facility Extension	520h	Work	16h	16h	16h	16h	8h			8h	8h
2	Approval to Bid	0h	Work									
3	◢ Technical Specification	152h	Work	16h	16h	16h	16h	8h			8h	8h
4	◢ Determine Installation Requirements	64h	Work	16h	16h	16h	16h					
	Project Manager	32h	Work	8h	8h	8h	8h					
	Systems Engineer	32h	Work	8h	8h	8h	8h					

- Highlight the rows on the left-hand side of your pane that you want to copy to the spreadsheet. This would be the Task name and Work in the picture above.

- Position the cursor under the spreadsheet heading and paste. The heading rows are copied and pasted with Microsoft Project 2010. These will paste as columns A and B in the picture below.

- Type the dates along the top row of your spreadsheet in the same format as the right-hand pane of Microsoft Project. As per cells C1 to K1 in the picture below.

- Select the data required to be pasted into Excel from the right-hand side of the pane in Microsoft Project by dragging. Ensure the top left-hand cell matches the first date typed in the spreadsheet.

- Paste into the spread sheet by placing the mouse in cell C2:

	A	B	C	D	E	F	G	H	I	J	K
1	Task Name	Work	07-Dec	08-Dec	09-Dec	10-Dec	11-Dec	12-Dec	13-Dec	14-Dec	15-Dec
2	Bid For Facility Extension	520h	16h	16h	16h	16h	8h			8h	8h
3	Approval to Bid	0h									
4	Technical Specification	152h	16h	16h	16h	16h	8h			8h	8h
5	Determine Installation	64h	16h	16h	16h	16h					
6	Project Manager	32h	8h	8h	8h	8h					
7	Systems Engineer	32h	8h	8h	8h	8h					

You may now create your S-Curves in the spreadsheet using its built-in Excel graphing function.

20.8.3 Creating S-Curves in Excel

Earned Value Performance Measurement (EVPM) data may be calculated in Microsoft Project (an example is shown at the end of the **UPDATING PROJECTS WITH RESOURCES** chapter) or in Excel. The table below is a simple example of EVPM data and this data may be displayed at any level such as at Project, Phase, WBS Node or task.

- The **Planned Value** is usually exported from Microsoft Project when the schedule is Baselined. This must be done before the schedule is progressed as this data may not be created from a program with progress.

- The **Earned Value** could be calculated either:
  - ➢ In Microsoft Project after the project has been updated and transferred to Excel, or
  - ➢ Calculated in Excel from the original Baseline Costs multiplied by the % Complete.

- The **Actual Costs** could be calculated either:
  - ➢ Outside Microsoft Project and entered into Excel, or
  - ➢ In Microsoft Project by entering the detail cost information against each task in Microsoft Project and exporting the rolled up cost into Excel each period.

- The **Estimate To Complete** is displayed in an EVPM graph and may be generated in Microsoft Project after the schedule, resources and costs have been revised and the Estimate To Complete exported to Excel at the required level. This is a very time consuming exercise if performed manually without a time sheet system.

	Jan	Feb	Mar	Apr	May	Jun	Jul	Aug
Planned or BCWS	$ -	$ 6,000	$ 15,000	$ 30,000	$ 43,000	$ 50,000	$ -	$ -
Earned or BCWP	$ -	$ 4,000	$ 10,000	$ 20,000	$ -	$ -	$ -	$ -
Actual or ACWP	$ -	$ 8,000	$ 20,000	$ 40,000	$ -	$ -	$ -	$ -
Estimate At Completion	$ -	$ -	$ -	$ 40,000	$ 55,000	$ 65,000	$ 70,000	$ 75,000

The Excel S-Curve Chart below may be created by highlighting the data in Excel, selecting a **Line** Chart Type and following the Wizard instructions. Histograms are created in a similar way by selecting a different Chart Type.

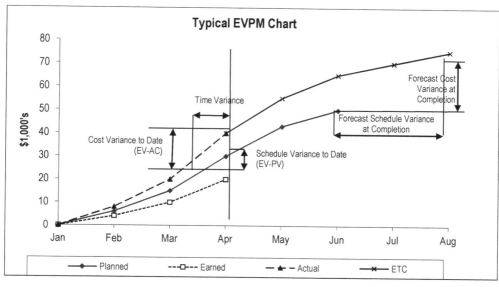

20.9 Resource Optimization

There are many techniques to optimize resource demand caused by:

- Overloading or peaks in resource requirements, or

- To resolve uneconomical use of resources due intermittent resource use resulting in additional costs required to demobilize and remobilize crews.

20.9.1 Resource Optimization through Leveling

After resource overloads or inefficient use of resources have been identified with Resource Graphs and Usage views, the schedule may have to be leveled to reduce peaks and troughs in resource demands. The process of leveling is defined as delaying tasks until resources become available. There are several methods of delaying tasks and thus leveling a schedule:

- **Turning off Automatic Calculation and Dragging Tasks**. This option does not maintain a Critical Path and reverts back to the original schedule when recalculated. This option should not be used when a contract requires a Critical Path schedule to be maintained.

- **Constraining Tasks**. A constraint may be applied to delay a task until the date that the resource becomes available from a higher priority task. This is not a recommended method because if the higher priority task is delayed the schedule may become unleveled. A delay Milestone should be added so the cause of the delay is easily seen when the plan is reviewed.

- **Sequencing Logic**. Relationships may be applied to tasks sharing the same resource(s) in the order of their priority. In this process a resource-driven Critical Path will be generated. If the first task in a chain is delayed then the chain of tasks will be delayed. But the schedule will not become unleveled and the Critical Path maintained. In this situation a successor task may be able to take place earlier and the logic will have to be manually edited.

- **Leveling Function**. The Leveling function will level resources by automatically delaying tasks without the need for Constraints or Logic and will find the optimum order for the tasks based on nominated parameters. Again, as this option does not maintain a Critical Path it should not be used when a contract requires a Critical Path schedule to be maintained. The Leveling function may be used to establish an optimum scheduling sequence and then Sequencing Logic applied to hold the leveled dates and to create a Critical Path.

20.9.2 Other Methods of Resolving Resource Peaks and Conflicts

Methods of resolving resourcing problems that are not strictly under the definition of leveling include:

- **Revising the Project Plan**. Revise a project plan to mitigate resource conflicts such as changing the order of work, contracting work out, prefabricating instead of site fabrication, etc.

- **Duration Change**. Increase the task duration and decreasing the resource requirements, so a 5-day task with 10 people could be extended to a 10-day task with 5 people.

 The user must be familiar with the Task Type as increasing durations will normally increase the work unless the Task is Fixed Work. Changing the duration of a Fixed Work task will not change the Work assigned to a task but the Author has found that the Units do not display correctly after a duration change to a Fixed Work task unless the Resource is Removed and reassigned.

- **Resource Substitution**. Substitute one resource with another available resource.

- **Increase Working Time by adjusting the calendar**. This will reduce the overall duration of the task may release the resource for other tasks earlier.

- **Splitting** tasks around peaks.

- Manually adjusting hours in **Usage** views.

20.9.3 Resource Leveling Function

Microsoft Project has a basic resource leveling function that is set up through the **Resource Leveling** form by selecting **Resource**, **Level** group, ⌐ **Leveling** Options.

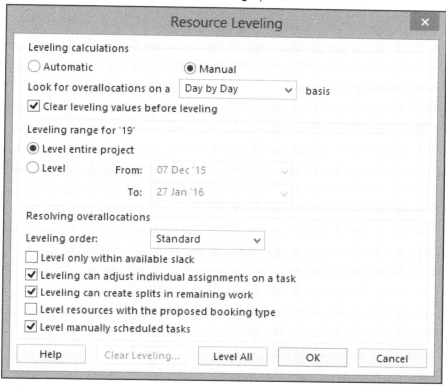

- **Leveling calculations**

  ➤ **Automatic/Manual**. When **Manual** is selected, the buttons in the **Resource Leveling** form are operational. When **Automatic** is selected, the schedule is always leveled according to the parameters selected in the **Resource Leveling** form; this is not a recommended option.

  ➤ **Look for overallocations on a ... basis** considers the average allocation in the time period selected and only levels when average exceeds the available, thus allowing for short periods of overloading.

  ➤ **Clear leveling values before leveling** un-levels the schedule and then re-levels.

- **Leveling range for...** allows for the whole project or a time portion of a schedule to be leveled, such as a Stage or Phase.

- **Resolving overallocations**

  ➤ **Leveling order:** allows three options for the order that tasks are prioritized for leveling. One option is the task **Priority** field which may be set in a column or in the **General** tab of the **Task Information** form. A **Priority** of 0 is the lowest priority and 1000 is the highest and tasks that must not be delayed should be set at 1000.

  ➤ **Level only within available slack** will not delay the end date of the project but will attempt to level the schedule in the available Total Slack (Total Float), but may result in overloading.

  ➤ **Leveling can adjust individual assignments on a task** allows one resource assigned to a task to be delayed in relation to others assigned to the same task, thus allowing other resources to work as scheduled. This will in effect increase the duration of a task and may result in a split task. This function works in conjunction with the **Level Assignments** field that may be displayed in a column. Individual tasks may or may not be allowed to have their resource assignments adjusted. This function has a similar effect to the Oracle Primavera P3 and SureTrak Task Type Independent which has been omitted from Oracle Primavera P6.

> **Leveling can create splits in remaining work;** allows leveling to split or not to split tasks during leveling.

> **Level resources with proposed booking type** will include these resources when leveling. The **Booking Type** may be set as **Proposed** or **Committed** in the resource sheet view.

> **Level manually scheduled tasks** will also include these tasks in the leveling calculations.

- Clear Leveling... allows the option of clearing leveling for selected task or all tasks and returns the schedule to the unleveled state.

20.9.4 Resource, Level group Ribbon Commands

This group of commands has the following options:

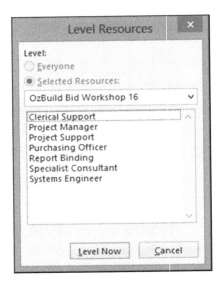

- → Level Selection levels two or more tasks that have been selected.

- Level Resources opens the Level Resources form allowing the selection of one or more resources for leveling.

- → Level All levels the project based on the parameters set in the Resource Leveling form.

- Clear Leveling un-levels the project.

- Next Overallocation scrolls through one overallocated task to another.

20.10 Workshop 17 – Resource Graphs and Tables

Background

We will create a copy of our current project file for this workshop, then use Usage Views and Graphs to isolate the resources that are over allocated and level the schedule.

Assignment

1. Save your **OzBuild Bid** project file and then resave as **OzBuild Leveling**.

2. Display the **Gantt Chart Inc Total and Neg Float** view in the top pane, the **Resource Usage** view in the bottom pane and adjust the timescale to show days only as per the picture below.

3. Click on **Select All button** to display all resources. The overloaded resources are highlighted in red.

4. Set the Timescale to daily. You may also wish to adjust the size % to fit more on the page. **NOTE:** The schedule below has been set to 60% size.

5. Right-click in the bottom pane right hand side and display both the Work and Overallocation resources. The overallocated resources are highlighted in red and the overallocated lines of data show by how much:

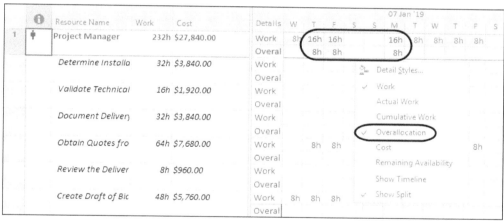

6. Display the **Resource Graph** form in the bottom pane and check the Histograms for the resources. Note again the overallocated resources:

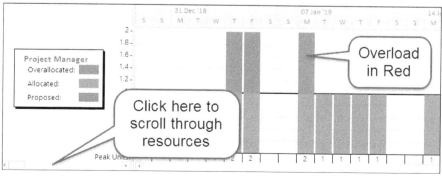

7. Set the Baseline and display the Baseline bars.

8. Now level the schedule, select **Resource**, **Leveling Options** to open the **Resource Leveling** form:

9. Set the options as per below and click on [Level All]:

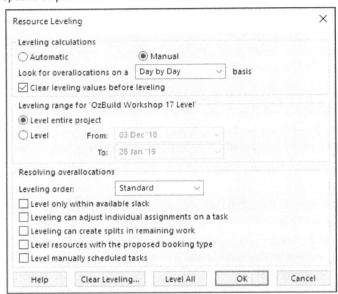

10. Your answer should look like this, with task 15 delayed and the end date also delayed:

 The author found that sometimes the leveling options did not calculate correctly unless they were changed and then changed back to the required settings.

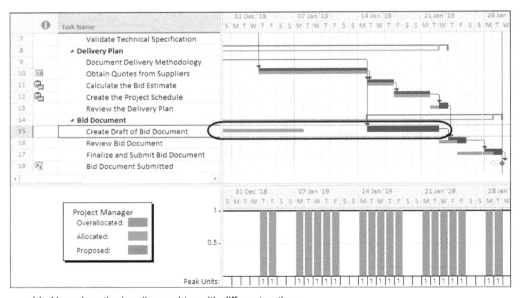

11. Now clear the leveling and try with different options.

21 UPDATING PROJECTS WITH RESOURCES

Updating a project with resources uses a number of features that are very complex and interactive. It is suggested that after you have read this chapter and before you work on a live project that you create a simple schedule with a couple of tasks and assign two or three resources against each task. Set the **Options** to reflect the way you want to enter the information and how you want Microsoft Project to calculate. Go through the updating process with dummy data and then check that the results are as you expected. The options to consider are:

- How is the measure of progress at the summary task level displayed? The summary **% Work** is based on hours of effort, which is more meaningful than the summary **% Complete**, which is based on durations. Or are you displaying the **Physical % Complete** which does not roll up to the summary tasks?

- Have you linked **% Complete** and **% Work** with the **File**, **Options**, **Schedule** tab, **Calculation options for this project:**, **Updating task status updates resource status** option? If unlinked, the **% Work** may be different from **% Complete**.

- Do you want Microsoft Project to calculate the resource **Actual Costs** with the option **File**, **Options**, **Schedule** tab, **Calculation options for this project:**, **Actual costs are always calculated by Project**?

- Are your tasks scheduled to start after a date using the **Project**, **Status** group, **Update Project**, **Reschedule uncompleted tasks to start after:** in conjunction with the **File**, **Options**, **Schedule** tab, **Scheduling options for this project:** section, **Split in-progress tasks** option?

Updating a project with resources takes place in two distinct steps:

- The dates are updated using the methods outlined in the **TRACKING PROGRESS** chapter, then
- The resources' hours and costs are updated.

This chapter covers the following topics:

- Understanding **Baseline Dates** (Target Dates), **Baseline Costs** (Budget) and **Baseline Work**.
- Understanding the **Status Date**, **Work After Date** and **Current Date** with respect to resources.
- Information required to update a Resourced Schedule.
- Updating Resources.
- Splitting Tasks.
- Summary Task Interim Baseline Calculation.

Microsoft Project does not provide the capability of producing multiple S-Curves. The data may be exported or copied and pasted to products such as Excel where S-Curves may be created.

 Microsoft Project 2013 calculates differently to earlier versions when the option **Actual costs are always calculated by Project** are unchecked when the activity is at 100%.

- In earlier versions the **Actual Cost** was unchanged, but could be manually changed that point on,
- In Microsoft Project 2013, the software changes the **Actual Costs** to zero when unchecked, which would normally be less desirable, and the **Actual Costs** must then be manually entered.

21.1 Understanding Baseline Dates, Duration, Costs and Hours

Baseline Dates are also known as Target Dates and are normally the original project Early Start and Early Finish dates. These are the dates against which project progress is measured.

Baseline Duration is the original planned duration of a task.

Baseline Costs are also known as Budgets and represent the original project cost estimate. These are the figures against which the expenditures and Cost at Completion (or Estimate at Completion) are measured.

Baseline Work is also known as Budgeted Quantity and represents the original estimate of the project quantities. These are the quantities against which the consumption of resources are measured.

The **Baseline Costs** and **Work** of resources are **NOT** automatically recorded in the Baseline fields of each resource as the resource is assigned to a task. All **Baseline** information **(Dates, Costs** and **Work)** is saved when the project or selected tasks have the Baseline set. Baseline data is saved at task and resource assignment level.

If resources have been assigned then the Baseline Costs and Work are recorded at the same time as the Baseline dates.

Setting the **Baseline Dates** is covered in the **TRACKING PROGRESS** chapter.

Baseline Dates may be displayed by using the following methods:

- Baseline Date Start and Finish columns, or
- Bar Chart Bars, or
- Dates on the Bars, or
- In forms such as the **Task Details** form.

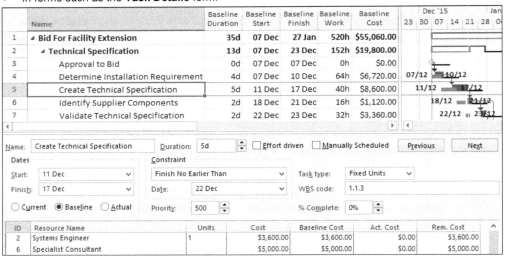

21.2   Understanding the Status Date

The **Status Date** is a standard scheduling term. It is also known as the **Data Date**, **Review Date**, **Status Date**, **As of Date**, **Report Date**, **Time Now** and **Update Date**.

- The **Status Date** is the date that divides the past from the future in the schedule. The **Status Date** is not normally in the future but is often in the recent past due to the time it may take to collect the information to update the schedule.

- **Actual Costs** and **Quantities/Hours** or **Actual Work** occur before the **Status Date**.

- **Costs** and **Quantities/Hours to Complete** or **Work to Complete** occur after the **Status Date**.

- **Remaining duration** is the duration required to complete a task. It is calculated forward from the **Data Date**.

Microsoft Project has four dates associated with updating a schedule. These are covered in the **TRACKING PROGRESS** chapter. In summary, these date fields are:

- **Current Date** – This date is set to the computer's system date, by default at 08:00hrs, each time a project file is opened. It is used to calculate **Earned Value** data when a **Status Date** is not set.

- **Status Date** – By default this is blank. After assigning this date, by default at 17:00hrs, it will not change (the **Current Date** does) when the project is saved or reopened at a later date. When set, this date overrides the **Current Date** for calculating **Earned Value** data.

- **Update work as completed through:** (date) – When a project is updated using the **Update Project** form (the project is updated as if it was progressing exactly according to plan), the **Status Date** is set to the same date as **Project Update Date** at 17:00hrs.

- **Reschedule uncompleted work to start after:** (date) – This function is used to move the **Incomplete Work** of **In-progress** tasks into the future.

  - ➢ **In-progress** tasks must be able to **Split** for this function to operate. The option to split tasks is found on the **File**, **Options**, **Schedule** tab.

  - ➢ The **Status Date** is **NOT** set to the **Reschedule uncompleted work to start after:** (date) when this function is used.

  - ➢ This function will not move the incomplete portions of tasks back in time when a task is completed ahead of schedule. This causes the task's **Remaining Duration** to occur sometime in the future and not immediately after the selected date. The example below shows a schedule with the **Reschedule uncompleted work to start after: (date)** set to 8 September. The top task's Remaining Duration commences on 9 September, the task is split and the Remaining Duration occurs immediately after 8 September. The lower task's Remaining Duration does not commence until 16 September. This is not realistic and manual intervention is required to correct this.

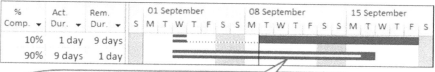

Scheduling completed work in the future is illogical so manual intervention is required.

NEITHER the **Current Date** nor the **Status Date** is used to calculate the **Early Finish** of an **In-progress** task when a schedule is calculated.

When **Automatic Scheduling** is disabled (e.g. the **File**, **Options**, **Schedule** tab, **Calculation** section **Calculate project after each edit is** not checked) then the schedule is recalculated using the **F9** key.

The end date of an in-progress task in Microsoft Project is normally calculated from the **Actual Start Date** plus the **Duration**. This is a different method of calculation than employed by other scheduling software, which normally calculates the end date of a task from a single **Data Date** plus the **Remaining Duration**. This ensures all incomplete work is in the future.

Users of Microsoft Project must spend more time checking the schedule to ensure that completed work is all in the past and incomplete work is in the future in relation to the **Status Date**.

There are a number of scheduling methods and functions available to achieve this situation but the software will not automatically make this happen.

21.3 Formatting the Current Date and Status Date Lines

To format the display of the **Current Date** and **Status Date** lines on the Bar Chart, select **Format**, **Format** group, **Gridlines…** to display the **Gridlines** form:

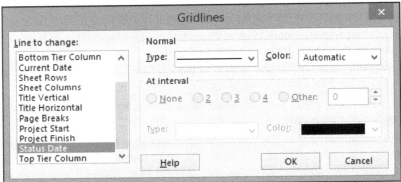

- These **Options** allow the selection of colors and line types for all sight lines as shown in the **Lines to Change** box on the left of the example above.
- All other sight lines may also be formatted in this form.

It is suggested that the **Current Date** Gridline be removed to avoid confusion with the **Status Date**.

The **Status Date** should be displayed in a color that is different to the relationship and grid lines so it stands out, for example solid black.

21.4   Information Required to Update a Resourced Schedule

A project schedule is usually updated at the end of a period, such as each day, week or month. One purpose of updating a schedule is to establish differences between the plan and the current schedule and if necessary, take action to bring the project back on track.

Microsoft Project may calculate task **Actual Costs** from the rates entered in the **Resource Table** or the costs may be entered manually. This option is set in the **File, Options, Schedule** tab. If the **Actual Costs** are to be calculated by Microsoft Project then the **Actual Costs** do not need to be collected. The following information is required to update a resourced schedule:

Tasks completed in the update period:

- **Actual Start** date of the task.
- **Actual Finish** date of the task.
- **Actual Costs** spent, **Actual Resource Hours** spent, and/or **Actual Material Quantities**.

Tasks commenced in the update period:

- **Actual Start** date of the task.
- **Remaining Duration** or **Expected Finish** date.
- **Actual Costs** and **Actual Resource Hours** and/or **Actual Material Quantities**.
- **Hours** or **Quantities** to complete. Costs to complete are always calculated using the resource rates in the Resource Table.
- **Stop** date and **Resume** date for tasks that have had their work suspended. These may be used for splitting tasks.

Tasks Not Commenced:

- Changes in logic or date constraints.
- Changes in estimated **Costs**, **Hours** or **Quantities**.
- Any additional tasks to represent new work.

The schedule may be updated once this information is collected.

You have the option of allowing Microsoft Project to calculate many of these fields from the **% Complete** by selecting the appropriate option found in the **File, Options, Schedule** tab.

21.5   Updating Dates and Percentage Complete

The schedule should be first updated as outlined in the **TRACKING PROGRESS** chapter. In summary, this is completed by entering:

- The **Actual Start** and **Actual Finish** dates of **Complete** tasks.
- The **Actual Start**, **% Complete**, **Remaining Duration** or **Expected Finish** of **In-progress** tasks.
- Adjust Logic and **Durations** of **Un-started** tasks.

Before you do this, you should set the **File, Options, Schedule** tab to ensure that the actual costs and hours calculate the way you want.

21.6   Updating Resources

There are many permutations available in the **File, Options** form for calculating resource data. Due to the number of resource options and numerous forms available in Microsoft Project, it is not feasible to document all the combinations available for resource calculation.

This book, therefore, outlines some typical scenarios and examples of entering the update data that you may want to try on your projects.

Material and **Cost** resources are updated in a similar way as **Work** resources and are not covered separately.

21.6.1   Updating Tasks with Fixed Costs Only

A project with fixed costs only is the simplest option for managing costs. The example below displays:

- A 10-day **Duration** task which is 50% Complete.

- A **Baseline Cost** of $1,000.00 which was created by an original **Fixed Cost** of $1,000.00 when the **Baseline** was set.

- A **Fixed Cost** of $800.00 representing a revised estimate that has reduced the original estimate by $200.00 to $800.00.

- An **Actual Cost** of $400.00 and **Remaining Cost** of $400.00, which are calculated from the 50 % **Complete**.

- The **Current Bar** (the upper bar in the picture below) shows that the task started 2 days late and is scheduled to end 2 days late as compared to the **Baseline Bar** (the black lower bar).

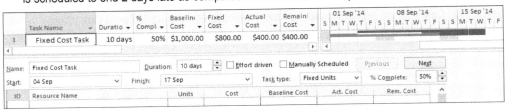

You will notice that the **Task Details** view with the **Resource Cost** details form does not show any resources or costs when **Fixed Costs** are used.

    The disadvantage of fixed costs is that they are always linked to the % Complete and an independent Actual Fixed Cost may not be entered. Microsoft Project 2007 introduced Cost Resources which has resolved this problem.

21.6.2   Forecasting Resource Hours

The next level of complexity usually occurs when a schedule is used for the management of resource hours but not costs.

One or more resources may be applied to a task and you may want to enter both the **Actual Work** and the **Remaining Work** independently. In this situation you will need to unlink % **Complete** and **Actual Work** with the <u>U</u>pdating task status updates resource status option in the **File**, **Options**, **Schedule** tab. Now the % **Work** field will be linked to the **Work**, **Actual Work** and **Remaining Work** fields and will now operate independently of the % **Complete** field.

The example below uses the **Task Details** form.

- A 10-day **Duration** task which is 40% Complete.

- The task's **Baseline Work** of 240 hours is from the addition of the two resource allocations when the **Baseline** was set.

  ➢ The Bid Manager was originally assigned at 100%, or full-time, giving a Baseline Work of 80 hours.

  ➢ The **Clerical Support** was originally assigned at 200%, or 2 full-time people, giving a **Baseline Work** of 160 hours.

- The task's **Actual Work** of 90 hours is the addition of the two resources' **Actual Work** of 20 hours and 70 hours.

- The task's **Remaining Work** of 220 hours is the addition of the **Remaining Work** of 70 hours and 150 hours.

- The **% Work** of 29% is calculated by dividing the **Actual Work** by the **Work.** This is different from the **% Complete**, which represents the elapsed time.

- Again, the **Current Bar** shows the task started two days late and is scheduled to end two days late.

- The **Actual Work** and **Remaining Work** were entered manually and have not been calculated by Microsoft Project from the **% Complete**.

- The duration of the task is still 10 days long because it is a **Fixed Duration** task. In the older version of the software below the **Units** have increased from 1 to 1.17 and 2 to 3.33 for the Bid Manager and Clerical Support, respectively, as the **Work** is now greater than the **Baseline Work** due to the increase in the number of hours required to complete the task. In this version of the software the values may be rounded and not represent the true value of the units.

21.6.3   Forecasting Resource Hours and Costs Form

The example below is similar to the previous example, but it now displays the **Resource Costs** (not the **Work**) calculated by Microsoft Project with % **Complete** and **Actual Work** unlinked.

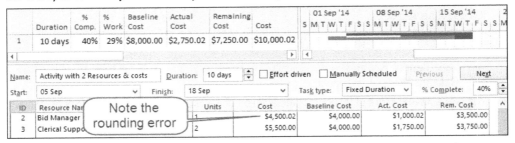

The next level of complexity occurs usually when a schedule is used for the management of resource hours and costs.

You may want to enter both the **Actual Work** and **Actual Costs** separately. For example, you may want to take the costs and hours from a timesheet and/or an accounting system that may have a different resource rate than your schedule. Input your data in the bottom pane of the window, which is known as the **Resource Details** form. This process is very time consuming and requires significant organizational commitment to succeed.

In this situation you will need to:

- Unlink % **Complete** and **Actual Work** with the **Updating task status updates resource status** option in the **File**, **Options**, **Schedule** tab, and

- Unlink the **Actual Work** and **Actual Costs** by disabling the **Actual costs are always calculated by Project** option in the **File**, **Options**, **Schedule** tab.

The example below shows how the costs are calculated with:

- **Work** and **Costs** unlinked, and

- Updated hours as per the previous example.

The **Actual Costs** have not been calculated by Microsoft Project and remain at zero. The **Remaining Costs** are based on the resources' **Standard Rate**. The **Actual Costs** should be entered by the user to update the schedule.

Now that the Actual Costs and Actual Work are unlinked, you may type in the **Actual Costs** per resource in the bottom pane of the window (i.e., the **Resource Details** form) without affecting the **Actual Work.**

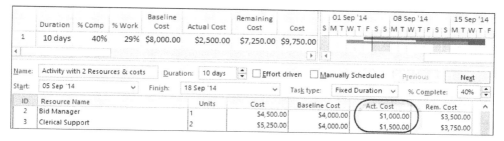

21.6.4 Using the Task Usage and Resource Usage Views

The **Task Usage** and **Resource Usage** forms allow the most flexibility when entering resource quantities and costs to used date, e.g. **Actual Work** and **Actual Cost**. The picture below is the **Task Usage** form from the OzBuild workshops showing the **Work**, **Actual Work** and **Baseline Work** rows. Additional rows of information may be obtained by right-clicking to display a menu.

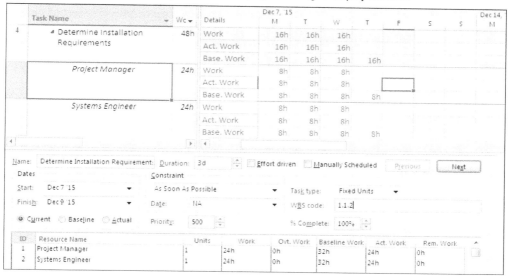

A timesheet-style view may be created with the **Resource Usage** view. This view may be printed or used by team members to directly update their hours to-date in the **Act. Work** (Actual Work) rows and their estimated hours in the **Work** rows. The advantage of this view is that all tasks are grouped together in one band under each name. The Project Manager's hours for Fri 11 Dec could be typed into the highlighted square above.

Calculated fields that may not have data entered into them are shaded, such as the task rows in the picture above.

21.7   Splitting Tasks

When the **Split in-progress tasks** option is enabled, a task may be **Split** by:

- Dragging the incomplete portion of a task in the bar chart, or

- Clicking on the ⊞ button and then moving your cursor over the point on the task bar where you want a split and dragging the task, or

- Using the **Project**, **Status** group, **Update Project**, **Reschedule uncompleted work to start after:** function, or

- Commencing a task before its predecessor finishes.

In the picture below the upper task was split using the ⊞ button and the lower task was split because it commenced before its predecessor.

A **Split** may be reversed by dragging the split portion back in the bar chart if there is no predecessor pushing the split out. The first task in the example below has been dragged back. The second task may not be dragged back due to the FS relationship. The relationship could be removed.

Once a task is **Split**, then the resources are not scheduled to work during the split period. This concept is demonstrated in the **RESOURCE OPTIMIZATION** chapter.

21.8   Summary Tasks and Earned Value

Actual Costs and **Work** may be summarized at any level in the same way as **Work** and **Costs**. The picture below is showing the **Earned Value** table. The costs have been summarized up to the Project Level:

	Task Name	Planned Value - PV (BCWS)	Earned Value - EV (BCWP)	AC (ACWP)	SV	CV	EAC	BAC	VAC
1	**Bid for Facility Extension**	$7,440.00	$7,908.00	$8,040.00	$468.00	-$132.00	$55,979.16	$55,060.00	-$919.16
2	**Technical Specificatio**	$7,440.00	$7,908.00	$8,040.00	$468.00	-$132.00	$20,130.53	$19,800.00	-$330.53
3	Approval to Bid	$0.00	$0.00	$0.00	$0.00	$0.00	$0.00	$0.00	$0.00
4	Determine Installat	$6,720.00	$6,720.00	$5,040.00	$0.00	$1,680.00	$5,040.00	$6,720.00	$1,680.00
5	Create Technical Sp	$720.00	$1,188.00	$3,000.00	$468.00	-$1,812.00	$21,717.17	$8,600.00	-$13,117.17
6	Identify Supplier Cc	$0.00	$0.00	$0.00	$0.00	$0.00	$1,120.00	$1,120.00	$0.00
7	Validate Technical S	$0.00	$0.00	$0.00	$0.00	$0.00	$3,360.00	$3,360.00	$0.00
8	**Delivery Plan**	$0.00	$0.00	$0.00	$0.00	$0.00	$21,520.00	$21,520.00	$0.00
14	**Bid Document**	$0.00	$0.00	$0.00	$0.00	$0.00	$13,740.00	$13,740.00	$0.00

The terminology below is used by a large number of companies and organizations. It provides standard terms to describe Earned Value calculations, which many people understand. Below are some of the terms that are in common use:

- AC or ACWP   Actual Cost or Actual Cost of Work Performed
- EV or BCWP   Earned Value or Budget Cost of Work Performed
- PV or BCWS   Planned Value or Budget Cost of Work Scheduled
- BAC   Budget At Completion
- C/SCSC   Cost/Schedule Control Systems Criteria (CS^2)
- CV   Cost Variance to date, BCWP – ACWP
- EAC   Estimate At Completion
- ETC Time   Estimate To Complete expressed in Time
- ETC   Estimate To Complete
- FAC $ Forecast at Completion
- FC CV   Forecast Cost Variance at Completion
                      (Budget – Forecast)
- FC SV   Forecast Schedule Variance at Completion
                      (Baseline End Date – Scheduled End Date)
- FTC CT   Forecast To Complete Calendar Time
- SV   Schedule Variance to date, BCWP – BCWS
- VAC   Variance At Completion

The **Physical % Complete** field may be used to calculate the Earned Value (Budget Cost of Work Performed) independently from the value in the task **% Complete** field. The **Physical % Complete** function is useful for measuring the progress of work that is not progressing linearly. This may be set for each individual task in the **General** tab of the **Task Information** form. The default for new tasks may be set with the [Earned Value...] button in the **General** tab of the **Options** form, where the Baseline to be used for the Earned Value calculations may be selected.

   The method that Microsoft Project uses to calculate the Earned Value data is documented in the Help file and should be read carefully, as different versions of Microsoft calculate these fields differently. Should different Earned Value calculations be required then Custom Data Fields should be considered as an alternative. The VAC column calculations should be checked and you will see the way Microsoft Project calculates the values and you may disagree with their method.

21.9 Workshop 18 - Updating a Resourced Schedule

Background

We need to update the tasks and resources.

Assignment

NOTE: If your settings are not exactly the same as the computer on which this exercise was undertaken or you enter data in a different order you may end up with different results.

1. Close the Leveling workshop and open your OzBuild with Resources project file and complete the following steps:

2. We will initially allow Microsoft Project to calculate costs and hours from the % Complete. Adjust the options using **File**, **Options**, **Schedule** tab, **Calculation options for this project:** to:

   ➢ **CHECK** the **Updating task status updates resource status** option. This will link % Complete and Actual Work. (This option also needs to be checked to allow Summary % Completes to be spread correctly to detail tasks.)

   ➢ **CHECK** the **Actual costs are always calculated by Microsoft Project** option. With this option checked the resource Actual Cost is calculated by Microsoft Project from the resource Work and Rates.

3. Ensure the **Split in-progress tasks** option in the **File**, **Options**, **Schedule** tab, **Scheduling options for this project:** section is UNCHECKED.

4. Save the Baseline using **Project**, **Schedule** group, **Set Baseline**.

5. Display the **Baseline** bars using **Format**, **Bar Styles**, **Baseline**.

6. Split the screen. Display the **Tracking Gantt View** in the upper screen and the **Task Details Form** with **Resource Work** details form in the Bottom Pane.

7. Format the Timescale to **Weeks 26 Jan '09** and **Days M,T,W,...** at 100% with a count of 1.

8. Check the **Status Date** is displayed as a solid black by selecting **Format**, **Format** group, **Gridlines**.

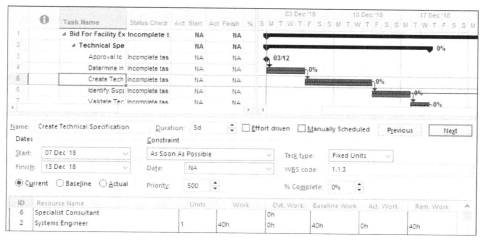

9. We are recording progress as of the end of the second week in December, so we will set the **Status Date** as Friday 7 Dec 2018. The time will default to the end of the work day:

> Update the project using **Project**, **Status** group, **Update Project**, as per the picture below:

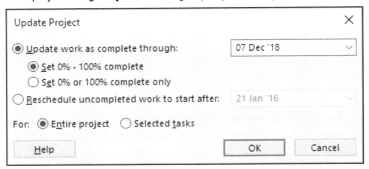

> Check the **Status Date** in the **Project**, **Properties** group, **Project Information** form. It should be 7 Dec 18 and the Status Date should be displayed on the Gantt Chart.

> The **Physical % Complete** has not been updated. This optional field may be used for entering the progress. It is not linked to the durations and may be used for calculating the Earned Value.

> Hide the **Physical % Complete** column if displayed.

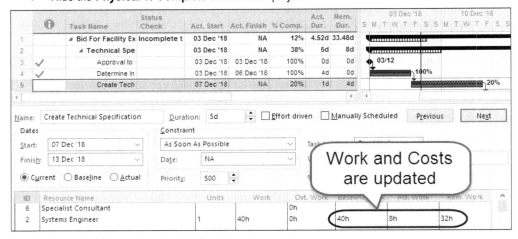

10. Update task **4 Determine Installation Requirements**

> Enter 48 hours in the **Actual Work** cell of task **4 Determine Installation Requirements**. **NOTE:** The Actual Finish date has now been removed and there is remaining work assigned to the task.

> Ensure the Remaining Work remains zero by typing zero against the resources in the Task Details Form. Press F9 if required.

> The duration shortens as the task is **Fixed Units** and the cost will change.

> The Actual Costs have been recalculated, see the **Resource Costs** tab in the bottom pane. Right click to display the menu to open this form.

> Also a new Actual Finish has been set without warning.

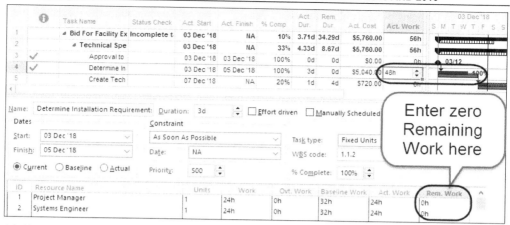

11. Updating **5 Create Technical Specification**

> Select task **5 Create Technical Specification** and look at the **Task Details Form**, **Work** and then **Cost** forms in the bottom pane, right click to display the menu to open these forms.

> Observe how the Systems Engineer's work and costs have been updated but the Specialists Consultant costs have not been updated. The Specialists Consultant cost could be manually updated if a cost had been incurred.

> Enter an Actual Start Date of 5 Dec 18.
> **NOTE:** There will be an error message in the **Status Check** column.

> Enter 60% Complete against task **5 Create Technical Specification** or click on the **Mark on Track** button. This should take the Actual Duration up to the Status Date and remove the **Status Check** column error message

> Task **6 Identify Supplier Components** should also start earlier as its predecessor is scheduled to finish earlier. Press F9 if required.

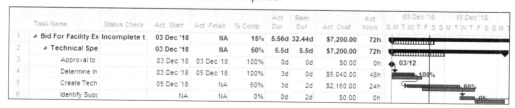

---

12. Enter 6d Remaining Duration against task **5 Create Technical Specification**, the **% Comp** should equal 33% and notice in the **Task Details** form that the remaining costs and remaining hours have increased.

> ➤ Press F9 to recalculate and your schedule should look like this:

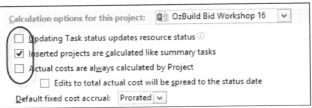

13. Now we will enter the hours and costs to date and hours to go. To prevent Microsoft Project from calculating the actual costs from the actual hours, open the **File**, **Options** form, **Schedule** tab, **Calculation options for this project** and:

> ➤ **UNCHECK** the **Updating task status updates resource status**. This option is to prevent the updated % Complete from calculating **Actual Work** and **Remaining Work** by unlinking the % Complete from the % Work fields.

> ➤ **UNCHECK** the **Actual costs are always calculated by Project**. This option is to prevent Microsoft Project from calculating the **Actual Costs** from the **Actual Work**.

 Microsoft Project 2013 and 2016 calculate differently to earlier versions when the option **Actual costs are always calculated by Project** is unchecked when the activity is 100%.

- In earlier versions the **Actual Cost** were unchanged, but could be manually changed that point on,

- In Microsoft Project 2013 and 2016, the software changes the **Actual Costs** to zero when unchecked, which would normally be less desirable, and then **Actual Costs** must then be manually re-entered.

14. Task **4 Determine Installation Requirements** will have the costs set to zero. Assign the following Actual Costs to the resources:

> ➤ $4,000.00 to the Project Manager and

> ➤ $3,000.00 to the Systems Engineer

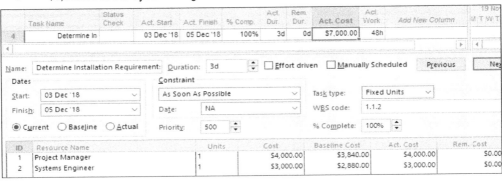

15. Update task **5 Create Technical Specification** with the following Actual Costs:

> ➤ **Actual Cost** of $3,000.00 against the **Systems Engineer**, and

> ➤ **Actual Cost** of $2,000.00 against the **Specialists Consultant**,

> ➤ Click the **OK** button to accept the changes. Press F9 if required.

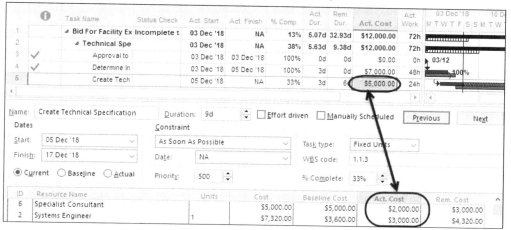

16. We will now update task **5 Create Technical Specification Remain Work**

> ➤ **Right**-click in the **Task Details Form** and display the **Work** tab, then

> ➤ Update task **5 Create Technical Specification**, with 72 hours of **Remaining Work** against the **Systems Engineer**,

> ➤ Apply the Cost details form in the bottom pane,

> ➤ The **Systems Engineer** Remaining Costs will recalculate based on the Remaining Work and Resource Unit Rate,

> ➤ Click the **Next** button or click out of the **Task** form to accept the changes.

NOTE: The task remaining duration extends to 9 days. This is because the task is **Fixed Units**, so the **Units** stayed the same and duration was extended. Your schedule should look like this:

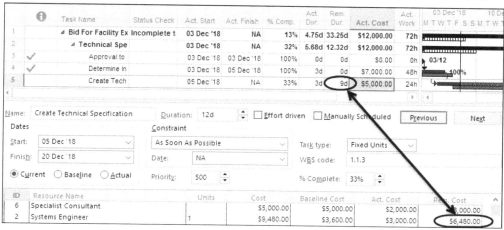

17. To see what the final forecast costs are, you may display the **Cost** table, the **Fixed Costs** and **Fixed Costs Accrual** columns have been hidden in the picture below:

	Task Name	Total Cost	Baseline	Variance	Actual	Remaining
1	⊿ Bid For Facility Extensio	$61,220.00	$55,060.00	$6,160.00	$12,000.00	$49,220.00
2	⊿ Technical Specificati	$25,960.00	$19,800.00	$6,160.00	$12,000.00	$13,960.00
3	Approval to Bid	$0.00	$0.00	$0.00	$0.00	$0.00
4	Determine Installatio	$7,000.00	$6,720.00	$280.00	$7,000.00	$0.00
5	Create Technical Sp	$14,480.00	$8,600.00	$5,880.00	$5,000.00	$9,480.00
6	Identify Supplier Cor	$1,120.00	$1,120.00	$0.00	$0.00	$1,120.00
7	Validate Technical S	$3,360.00	$3,360.00	$0.00	$0.00	$3,360.00
8	⊿ Delivery Plan	$21,520.00	$21,520.00	$0.00	$0.00	$21,520.00
9	Document Delivery I	$3,840.00	$3,840.00	$0.00	$0.00	$3,840.00
10	Obtain Quotes from	$12,160.00	$12,160.00	$0.00	$0.00	$12,160.00
11	Calculate the Bid Es	$1,920.00	$1,920.00	$0.00	$0.00	$1,920.00
12	Create the Project S	$1,920.00	$1,920.00	$0.00	$0.00	$1,920.00
13	Review the Delivery	$1,680.00	$1,680.00	$0.00	$0.00	$1,680.00
14	⊿ Bid Document	$13,740.00	$13,740.00	$0.00	$0.00	$13,740.00
15	Create Draft of Bid I	$8,160.00	$8,160.00	$0.00	$0.00	$8,160.00
16	Review Bid Docume	$3,360.00	$3,360.00	$0.00	$0.00	$3,360.00
17	Finalize and Submit	$2,220.00	$2,220.00	$0.00	$0.00	$2,220.00
18	Bid Document Subm	$0.00	$0.00	$0.00	$0.00	$0.00

18. To see how the consumption of hours is going you may display to **Work** table:

	Task Name	Work	Baseline	Variance	Actual	Remaining	% W. Comp.
1	⊿ Bid For Facility Extensio	560h	520h	40h	72h	488h	13%
2	⊿ Technical Specificati	192h	152h	40h	72h	120h	38%
3	Approval to Bid	0h	0h	0h	0h	0h	100%
4	Determine Installatio	48h	64h	-16h	48h	0h	100%
5	Create Technical Sp	96h	40h	56h	24h	72h	25%
6	Identify Supplier Cor	16h	16h	0h	0h	16h	0%
7	Validate Technical S	32h	32h	0h	0h	32h	0%
8	⊿ Delivery Plan	224h	224h	0h	0h	224h	0%
9	Document Delivery I	32h	32h	0h	0h	32h	0%
10	Obtain Quotes from	128h	128h	0h	0h	128h	0%
11	Calculate the Bid Es	24h	24h	0h	0h	24h	0%
12	Create the Project S	24h	24h	0h	0h	24h	0%
13	Review the Delivery	16h	16h	0h	0h	16h	0%
14	⊿ Bid Document	144h	144h	0h	0h	144h	0%
15	Create Draft of Bid I	96h	96h	0h	0h	96h	0%
16	Review Bid Docume	32h	32h	0h	0h	32h	0%
17	Finalize and Submit	16h	16h	0h	0h	16h	0%
18	Bid Document Subm	0h	0h	0h	0h	0h	0%

19. Now apply the **Earned Value** table, you may need to select the **More Tables** option to find this table. The VAC column does not calculate vertically. Microsoft Project calculates the EAC horizontally using the formula EAC = ACWP + (Baseline cost X - BCWP) / CPI which is not the same value as calculating vertically:

	Task Name	Planned Value - PV (BCWS)	Earned Value - EV (BCWP)	AC (ACWP)	SV	CV	EAC	BAC	VAC
1	⊿ Bid For Facility Extensio	$7,440.00	$7,908.00	$10,000.00	$468.00	-$2,092.00	$69,625.70	$55,060.00	-$14,565.70
2	⊿ Technical Specificati	$7,440.00	$7,908.00	$10,000.00	$468.00	-$2,092.00	$25,037.94	$19,800.00	-$5,237.94
3	Approval to Bid	$0.00	$0.00	$0.00	$0.00	$0.00	$0.00	$0.00	$0.00
4	Determine Installatio	$6,720.00	$6,720.00	$7,000.00	$0.00	-$280.00	$7,000.00	$6,720.00	-$280.00
5	Create Technical Sp	$720.00	$1,188.00	$3,000.00	$468.00	-$1,812.00	$21,717.17	$8,600.00	-$13,117.17
6	Identify Supplier Cor	$0.00	$0.00	$0.00	$0.00	$0.00	$1,120.00	$1,120.00	$0.00
7	Validate Technical S	$0.00	$0.00	$0.00	$0.00	$0.00	$3,360.00	$3,360.00	$0.00
8	⊿ Delivery Plan	$0.00	$0.00	$0.00	$0.00	$0.00	$21,520.00	$21,520.00	$0.00
9	Document Delivery I	$0.00	$0.00	$0.00	$0.00	$0.00	$3,840.00	$3,840.00	$0.00
10	Obtain Quotes from	$0.00	$0.00	$0.00	$0.00	$0.00	$12,160.00	$12,160.00	$0.00
11	Calculate the Bid Es	$0.00	$0.00	$0.00	$0.00	$0.00	$1,920.00	$1,920.00	$0.00
12	Create the Project S	$0.00	$0.00	$0.00	$0.00	$0.00	$1,920.00	$1,920.00	$0.00
13	Review the Delivery	$0.00	$0.00	$0.00	$0.00	$0.00	$1,680.00	$1,680.00	$0.00
14	⊿ Bid Document	$0.00	$0.00	$0.00	$0.00	$0.00	$13,740.00	$13,740.00	$0.00
15	Create Draft of Bid I	$0.00	$0.00	$0.00	$0.00	$0.00	$8,160.00	$8,160.00	$0.00
16	Review Bid Docume	$0.00	$0.00	$0.00	$0.00	$0.00	$3,360.00	$3,360.00	$0.00
17	Finalize and Submit	$0.00	$0.00	$0.00	$0.00	$0.00	$2,220.00	$2,220.00	$0.00
18	Bid Document Subm	$0.00	$0.00	$0.00	$0.00	$0.00	$0.00	$0.00	$0.00

22 PROJECT OPTIONS

The Options define the rules under which Microsoft Project operates.

The traditional **Options** form has been revamped in Microsoft Project 2010 and these are now under the **File**, **Options** command which opens the **Project Options** form.

The **Options** forms allow you to decide how Microsoft Project calculates and displays information. Most of the options are self-explanatory.

There are many options that are not essential for scheduling. Unfortunately, it is often difficult for new users to determine which are important and which are not, which can lead to confusion and potentially some scheduling errors. This chapter will explain most of the functions identified on each Options form tab and indicate the important options.

- **General**   This is used for editing the Default Date format, Default View and color.

- **Display**   Includes the selection of the currency and options for indicators buttons.

- **Schedule**   This form has a number of sections controlling how the software calculates and **SHOULD** be understood.

- **Proofing**   This form is used to decide how the spell checker works.

- **Language**   This is where the language options may be edited.

- **Advanced**   This new form has display options that are intended to assist the user and provide some further advanced scheduling options. These options **SHOULD** be understood.

- **Customize Ribbon**   Commands may be added to and removed from the Ribbon.

- **Quick Access Toolbar**   Commands may be added to and removed from the Quick Access Toolbar.

- **Add-Ins**   This is where Add-Ins may be viewed and managed:

- **Trust Center**   Allows the control of privacy and macro security.

There are more Options in the Professional version of the software and later versions tend to also have more options but the main ones do not change

Some options apply to the current project only, some to all projects and some to new projects:

- Those that apply to the current project or new project have a drop-down box where this selection may be made.

- Those that apply to all projects do not have any options to select a project.

 It is important to understand what the options do and to carefully inspect the options of schedules sent to you.

22.1   General

This is used for editing some global defaults such as the Default Date format, Default View and color scheme:

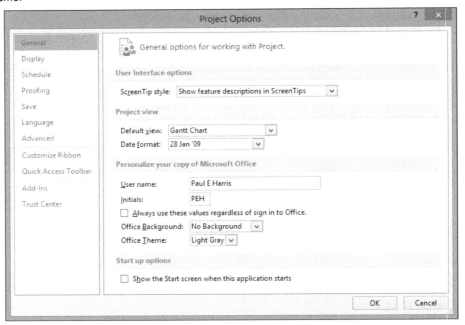

22.1.1   User Interface Options

- **ScreenTip style:** shows the name and a brief description of a command when the mouse is hovered over a button; it is recommended this be left on.

22.1.2   Project view

- **Default view:** selects the View applied when a new project is created.. A default load of Microsoft has this set as **Gantt with Timeline** but should be change to **Gantt Chart** to provide more work area.

- **Date format:** provides a selection of date formats from the drop-down box for the display of data in columns. This format is applied to all projects and all views.
  - ➢ The date format will be displayed according to a combination of your system default settings and the Microsoft Project Options settings. You may adjust your date format under the system **Control Panel**, **Regional and Language Options**. Changing the Control Panel setting will change the options available in the Microsoft Project **Options** form.
  - ➢ The dates on the bars are formatted in the **Format**, **Format** group, **Layout** form.
  - ➢ This date format may be overridden by defining specific dates for each Table in the **Table Definition** form.

- Should you wish to display your times in 24-hour format, e.g. 17:00 instead of 5:00pm, which prevents confusion and makes it easier to enter times into fields then:
  - ➢ Select from the Windows menu **Settings**, **Control Panel**, **Regional Options** or **Region & Language** tab, depending on your operating system.
  - ➢ Select **Customize**, select the **Time** tab and
  - ➢ Change the time format from **h:mm:ss tt** to **HH:mm:ss** or **H:mm:ss**.

Some versions of Windows may have slightly different commands and this setting will affect many other applications, so times in Excel and Outlook etc. will be in the assigned format.

There is often confusion on international projects between the numerical US date style, mmddyy and the numerical European date style, ddmmyy. For example, in the United States 020719 is read as 07 Feb '19 and in many other countries as 02 Jul '19. Consider always adopting either the ddmmyy style, **06 Jan '19** or mmmddyy style, **Jan 06 '19** for all your plans when multiple countries are involved in a project.

22.1.3   Personalize your copy of Microsoft Office

- The **User name**: and Initials: may be called up in print headers and footers and sets the default Author in the **Properties** form.

- **Office Background:** and **Office Theme**: allows the selection of background colors which mainly affect the print out of Timescale and Column headings.

22.1.4   Start up options

Start up options: unlike earlier versions of Microsoft Project, a blank project titled **Project 1** is not automatically created, so you will either need to:

- Create a new project or open an existing project before you may start working, or

- If you wish the software to start up in the Gantt Chart View with a blank project titled **Project 1** created (as in earlier version of Microsoft Project) then you will need to select **File**, **Options**, **General**, **Start up options** and uncheck the **S̲how the Start screen when this application starts**:

22.2   Display

This form includes the selection of the currency and options for indicators buttons:

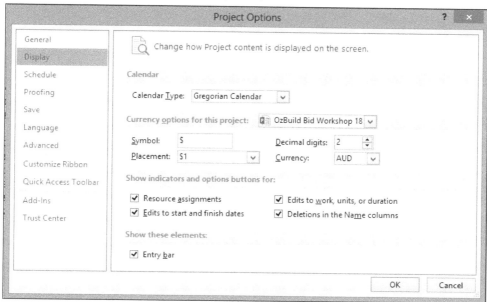

22.2.1 Calendar

Calendar Type: allows the selection of different calendars if the appropriate character set has been installed. This was new to Microsoft Project 2007.

22.2.2 Currency options for this project:

The **Currency:** option was new to Microsoft Project 2007 and allows the specification of the currency used in the schedule.

After a project has been created this option allows you to specify the:

- Currency sign by inserting a sign into the box next to **Symbol:**,

- Placement of the currency sign (before or after the value) as selected from the **Placement:** box, and

- The number of decimal places displayed from the **Decimal digits:** box.

 The default is set by the operating system settings and may not be set for all projects unless a Template is used which has a different currency assigned.

22.2.3 Show indicators and options buttons for:

The Indicator buttons give advice to the scheduler when certain functions are used which have more than one potential outcome.

- Below is an example of how the **Show indicators and Option buttons for:** options work when a task description is deleted:

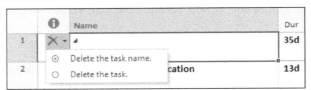

22.2.4 Show these elements

- The **Entry Bar** is at the top of the columns, data may be edited here when it is displayed:

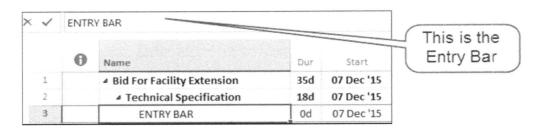

22.3　Schedule

This form has a number of sections controlling how the software calculates and **SHOULD** be understood.

22.3.1　Calendar options for this project:

The title of these options does not clearly indicate their functionality. Options for **Hours per day:**, **Hours per week:** and **Days per month: SHOULD** be understood, especially if multiple Task calendars are used.

- This heading allows the setting of the default for **All New Projects** or for individual projects. This sets the current calendar settings as the default in the Global.mpt project and is used thereafter as the calendar settings in all new projects.

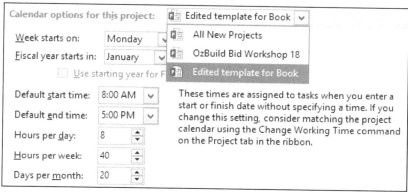

- **Week starts on:** – Sets the first day of the week. This is the day of the week that is displayed in the timescale. This setting affects the display of the calendar in the Change Working Time form. **NOTE:** This is best set to Monday and not Sunday so the date displayed in a weekly timescale is a working day.

- **Fiscal year starts in:** – A month other than January may be selected as the first month of the fiscal year for companies that want to schedule projects using their financial years.

- **Use starting year for FY numbering** – This option becomes available when a month other than January is selected as a **Fiscal year starts in:**. The year in the timescale may be assigned as the year that the fiscal starts or the year that the fiscal year finishes.

- **Default start time:** – This is the time of day that tasks are scheduled to start when a start date constraint (without a time) is applied. It is also the default time for an Actual Start when an Actual Start (without a time) date is entered.

- **Default end time:** – This is the time of day that tasks are scheduled to finish when a finish date constraint (without a time) is entered. It is also the default time for an Actual Finish when an Actual Finish date (without a time) is entered.

 It is essential to match both the **Default start time:** and **Default end time:** with your project calendar's normal start and finish times. For example, when the **Default start time:** is set earlier than the calendar normal start time, a task set with a constraint without a time will appear to finish one day later than scheduled. Therefore a one-day task will span two days and a two-day task, three days.

The following options need to be understood as they affect how summary durations are displayed. **NOTE: THESE OPTIONS ARE IMPORTANT**. Microsoft Project effectively calculates in hours. Task durations may be displayed in days, weeks and months. These summarized durations are calculated based on the parameters set in **Options**, **Calendar**. These options work fine when all project calendars are based on the same number of work hours per day. When tasks are scheduled with calendars that do not conform to the **Options**, **Calendar** settings (e.g., when the **Options**, **Calendar** settings are set for 8 hours per day and there are tasks scheduled on a 24-hour per day calendar), the results often create confusion for new users.

- **Hours per day:** – This setting is used to convert the duration in hours to the displayed value of the duration in days. For example, a task entered as a 3-day duration and assigned a 24-hour per day calendar will be displayed in the bar chart as 1-day elapsed duration. This bar chart duration is 1/3 of the duration value since each day is an 8-hour equivalent on a 24-hour calendar and the output is not logical. See Task 1 below.

- **Hours per week:** – This setting is used to convert the calendar hours to the displayed duration in weeks. For example, a task entered as 2 weeks (10 working days) and assigned a 24-hour per day calendar is displayed with a 3-day and 6-hour duration in the bar chart. This is 1/3 of the "normal" duration and again not logical. See Task 2 below.

- **Days per month:** – This option is used to convert the displayed task duration from days to months. For example, a 0.5-month task (10 working days) on a 24-hour per day calendar is displayed as a 3-day and 6-hour duration in the bar chart and again is not logical. See Task 3 below.

	Duration	Task Calendar	Mon 01 Sep				Tue 02 Sep				Wed 03 Sep				Thu 04 Sep			
			0	6	12	18	0	6	12	18	0	6	12	18	0	6	12	18
1	3 days	24 Hours																
2	2 wks	24 Hours																
3	0.5 mons	24 Hours																

 All the durations displayed in the picture are misleading. It is important that this situation be avoided.

Other scheduling software also exhibits the time conversion problem when using multi-calendars. This display can lead to a great deal of confusion. To avoid confusion when using multi-calendars, it is suggested that you only display durations in hours.

This issue is discussed in depth in para 24.1. Refer to this chapter should multiple calendars be considered for a project schedule.

22.3.2   Schedule

- **Show scheduling messages** – Uncheck this box to prevent the software from displaying advice on scheduling.

 The advice given by Microsoft is often confusing and opposite to what the user is trying to achieve. For example when a Finish No Later than Constraint is being set the software will advise a Finish No Earlier than Constraint, the opposite to what is trying to be achieved.

- **Show assignment unit as a:** – The options are **Percentage** or **Decimal**. You have the option of assigning and displaying resource assignments as either a percentage, such as 50%, or a decimal, such as 0.5. Both these values would assign a person to work half time on that task. When four resources are assigned to a task this would be 400% as **Percentage** or 4 as **Decimal**.

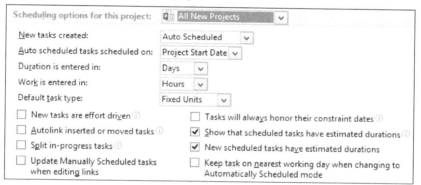

22.3.3   Scheduling options for this project:

The options below affect how your schedules calculate and **SHOULD** be understood. They may be applied to all new projects or to the current project:

Microsoft has introduced this new task attribute titled **Task Scheduling Mode** in Microsoft Project 2010. In earlier versions of Microsoft Project every task was set to the equivalent of the new **Auto Scheduled** setting in Microsoft Project 2010. In this mode all tasks' Start and Finish dates are calculated with respect to the Project Start (or Finish) date, their Durations, Relationships, Constraints, Calendars and Options. When a task is set to **Manually Scheduled** it does not need a predecessor or constraint to hold the Start date and therefore will not move when the project is scheduled.

- **New tasks created:** When new tasks are created they may be set with the **Task Mode** as either:

  ➤ **Manually Scheduled**, or

  ➤ **Auto Scheduled**.

  The attribute assigned to each task may be changed at a later date using the **Task Mode** column, or the **Task Details** form or the **Task Information** form. This option only sets the default for new tasks. This topic is covered in detail in para 24.3.1 and will not be covered in detail here.

 To produce a Critical Path schedule, tasks have to be set to **Auto Scheduled**.

- **Auto scheduled tasks scheduled on:** – When adding new tasks, you have the option of the new task either:

  - Scheduled to **Start On Project Start Date**, the task will be scheduled **As Soon As Possible**, or
  - **Start on Current Date**, the task will be assigned (without any warning dialog box) a **Start No Earlier Constraint** equal to the current date. **NOTE:** To have a task assigned a constraint without due reason is not considered good scheduling practice and should be avoided, so this option should not be selected.

- **Duration is entered in:** – You should select the units you will most frequently use for your task duration units. When "days" are selected, you will only need to type "**5**" in order to enter a 5-day task duration. You will have to enter "**5w**" for a task that is 5 weeks long or "**5h**" for a task that is 5 hours long.

- **Work is entered in:** – This is similar to the **Duration is entered in:** function, but applies to assigning work to a task. When hours are selected, you will only need to type "**48**" to enter 48 hours of work. If 5 days of work are to be assigned then "**5d**" needs to be entered.

 The next two subjects, **Default task type:** and **Effort-Driven**, are **EXTREMELY** important for understanding how resources are used and applied.

- **Default task type:** – There is a relationship between the **Duration** of a task, the **Work** (the number of hours required to complete a task) and the **Units** (per Time Period, the rate of doing the work or the number of people working on the task). The relationship is:

  Duration x Units (per time period in Hours or Days) **= Work** (Units)

  The three options for the **Default task type:** are:

  - **Fixed Duration** – The **Duration** will stay constant if either **Units** (per time period) or **Work** is changed. If you change the **Duration**, then **Work** changes. This is the **RECOMMENDED** option for new users and resourced schedules as the duration will not change as resources are manipulated.
  - **Fixed Units** – The **Units** stay constant if either **Duration** or **Work** is changed. If you change the **Units**, then **Duration** changes. This is the **RECOMMENDED** option for new users and unresourced schedules as progress and % Complete is displayed better.
  - **Fixed Work** – The **Work** stays constant if either **Duration** or **Units per Time Period** is changed. Your estimate will not change when you change **Duration** or **Units**. If you change the **Work**, then **Duration** changes.

Duration type	Change Duration	Change Units/Time	Change Work
Fixed Duration	Work Changes	Work	Units/Time
Fixed Units	Work Changes	Duration	Duration
Fixed Work	Units/Time	Duration	Duration

- **New tasks are effort driven** – An **Effort-Driven** task keeps the total work constant as resources are added and removed from a task. It treats each resource as interchangeable. When a task is not **Effort-Driven** then the addition of resources to a task will increase the total work assigned to a task. This option is not available with a **Fixed Work** task because if it was available then it would not be fixed work. **NOTE:** New users should consider **NOT** checking this option, so resources may be added independently without affecting the other resources work assignment.

- **Autolink inserted or moved tasks** – With this box checked, new or moved tasks are automatically linked to the tasks above and below in the new or inserted location. Moved tasks that had relationships to tasks immediately above and below in the old location are deleted and the original predecessors and successors are now joined with an FS relationship. This function may confuse you since the logic is changed automatically without warning, when you add or move tasks. **NOTE:** It is suggested that this option **NEVER** be switched on as dragging a task to a new location may completely change the logic of a schedule.

- **Split in-progress tasks:** – This option allows the splitting of in-progress tasks. This option **MUST** be checked for the following functions to operate:

  1. **Project**, **Status Group**, **Update Project** form, **Update work as completed through**,

  2. **Project**, **Status Group**, **Update Project** form, **Reschedule uncompleted work to start after:**, and

  3. **File**, **Options**, **Advanced** tab, **Calculation options for this project:** section allowing incomplete or completed parts of tasks to be moved to the **Status Date**.

  4. You may wish to turn off **Auto Calculation** in the **File**, **Options**, **Schedule**, **Calculation** section before assigning an **Actual Start** as this will create a split at the start of the task when turned on and an Actual Start applied You will then need to press F9 to reschedule after updating the tasks.

  This function will make some tasks split when they actually start before their predecessors. This can get confusing for new users when tasks generate splits. This function also creates a more conservative schedule normally taking longer to complete the tasks. The author suggests that new users turn this option off until they have some experience with the software. This topic is covered in detail in para 16.3.4 Calculating the Early Finish Date of an In-Progress Task.

- **Update Manually Scheduled tasks when editing links**

  When two **Manually Scheduled** tasks are linked they acknowledge the relationship and will reschedule when:

  ➢ The **File**, **Options**, **Schedule** tab, **Scheduling options for this project:** section, **Update Manually Scheduled tasks when editing links** is checked, and

  ➢ A relationship is added or edited.

- **Tasks will always honor their constraint dates**

  This option will make all constraints override relationships. For example, a task with a **Must Start On** constraint, which is prior to a predecessor's Finish Date, will have an Early Start on the constraint date and not the scheduled date. (This is similar to converting all Primavera software **Must Start On** constraints to **Mandatory** constraints.) When checked, the **Total Slack** may not calculate as the difference between Late Start and Early Start. Examine the following two examples with the option box checked and unchecked:

  ➢ **Tasks will always honor their constraint dates** option box checked.

Start	Finish	Late Finish	Total Slack	Constraint Date	Constraint Type	Aug '14 T W T F S S	25 Aug '14 M T W T F S S
12 Aug	28 Aug	25 Aug	-3 days	NA	As Soon As Possible		
29 Aug	29 Aug	26 Aug	-3 days	NA	As Soon As Possible		
26 Aug	26 Aug	26 Aug	-3 days	26 Aug	Finish No Later Than		

 The third task starts before the predecessor finishes and the Total Slack of the second task is calculated at –3 days, which is not the difference between the early and late dates. Thus the Total Float calculation is also incorrect.

➤ **Tasks will always honor their constraint dates** option box NOT checked.

Start	Finish	Late Finish	Total Slack	Constraint Date	Constraint Type		
12 Aug	28 Aug	25 Aug	-3 days	NA	As Soon As Possible		
29 Aug	29 Aug	26 Aug	-3 days	NA	As Soon As Possible		
29 Aug	29 Aug	26 Aug	-3 days	26 Aug	Finish No Later Than		

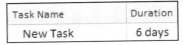

 It is suggested that this option is **NEVER** switched on, unless you are very experienced, as the schedule may appear to be achievable when it is not.

- **Estimated Durations** – These two options do not affect the calculation of projects. A new task is assigned an estimated duration. Once a duration is entered by the user, this assignment is removed. The **General** tab on the **Task** form has a check box which establishes when a task has an estimated duration.

  ➤ **Show that scheduled tasks have estimated durations** – A task with an estimated duration is flagged with a "**?**" after the duration in the duration column when this option is checked.

  ➤ **New scheduled tasks have estimated durations** – When a task is added, it will have the estimated check box checked.

Before entering a Duration   and after entering a Duration

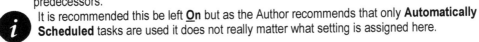

Task Name	Duration	Task Name	Duration
New Task	1 day?	New Task	6 days

 Summary tasks are marked as estimated when one or more **Detailed tasks** associated with it are marked as Estimated.

- **Keep Tasks on nearest working day when changing to Automatically Scheduled mode.**

  When a task is **Manually Scheduled** and changed to **Automatically Scheduled** this option:

  ➤ When checked set a **Start No Earlier** constraint on the task so it does not move forward in time if it has no predecessors, or

  ➤ When unchecked the task will not have a constraint set and will move forward if it has no predecessors.

   It is recommended this be left **On** but as the Author recommends that only **Automatically Scheduled** tasks are used it does not really matter what setting is assigned here.

22.3.4 Schedule Alerts

This allows the turning on or off of advice from Microsoft Project:

Schedule Alerts Options: 📇 All New Projects ▽

☐ Show task schedule warnings
☐ Show task schedule suggestions

⚠ It is suggested that these types of warnings are either read very carefully or disabled as they often encourage schedulers to do the wrong thing, such as assign a Finish No Earlier Than Constraint when the scheduler requires a Finish No Later Than Constraint.

22.3.5 Calculation

Select one of the following two options, which affect the calculation of **Automatically Scheduled** tasks for **ALL** projects that are open:

- **On** – This option recalculates the schedule every time a change is made to data. This is the default option.

- **Off** – This option requires you to press the **F9 Key** to recalculate the schedule.

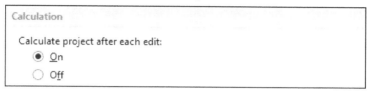

 It is recommended this be left **On** as Microsoft Project calculates very quickly.

22.3.6 Calculation Options for this project:

These options may be assigned to New Projects and selected projects.

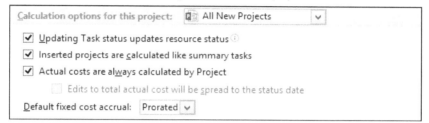

- **Updating task status updates resource status** – This option links % **Complete** to % **Work** and in turn % **Work** is linked to **Actual Work**. Therefore when this option is checked:

  Actual Work = % **Work x Work** and

  Remaining Work = **Work – Actual Work**

When unchecked the % **Complete** and % **Work** may have different values and % **Work** or **Actual Work** will have to be entered separately from the % **Complete**.

The example below shows the calculations with this option checked. If you uncheck this option, you may enter Actual Work and Remaining Work independently of the % **Complete**. In the example below the % Work is 50% and % Complete is 20%.

A summary task % **Work** is calculated based on the sum of **Actual Work** divided by the sum of the **Work** and gives an indication of progress based on hours of effort.

Understanding Summary Duration % Complete

A summary task **% Complete** is calculated from the sum of **Actual Durations** of all the **Detailed Tasks** divided by the sum of the **Durations**.

This option also determines the manner in which a % Complete that is assigned to a Parent Task gets spread to detail tasks. See the following two examples where 60% is applied to the summary task entitled "Research":

- With the **Updating task status updates resource status** CHECKED.

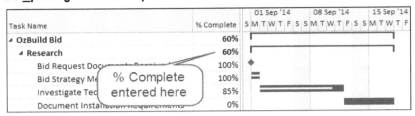

- With the **Updating task status updates resource status** UNCHECKED.

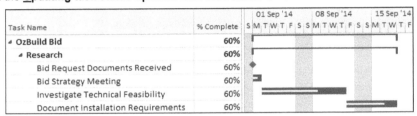

- **Inserted projects are calculated like summary tasks**

  A project may be inserted like a sub-project using the **Project**, **Insert**, **Subproject** command.

  ➤ When checked, an inserted project task's Total Float is calculated to the end of all Projects like a group of related projects.

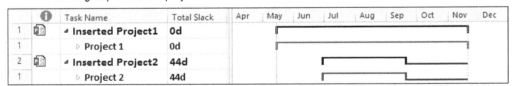

  ➤ When unchecked, each Project's Total Float should be calculated to the end of that Project only.

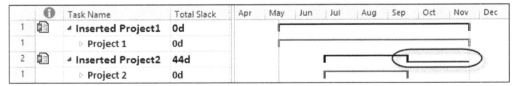

 The Inserted Project Tasks calculate correctly but the Inserted Project Summary bars do not, as per the picture above. The float bar on the summary task may be removed manually using the **Bars** form.

When there are several Microsoft Project projects inserted into a Master project file, they may be treated as:

- Part of the same project with this option checked and the float will be calculated to the end of the last activity of all projects, then the option should be switched on, or

---

- A group of unrelated projects where the float of each project is calculated independently, then the option should be switched off.

- **Actual costs are always calculated by Project** – With this option checked, the resource **Actual Cost** is calculated by Microsoft Project from the resource **Work** and **Rates**. When unchecked, you have to directly enter the **Actual Cost** and **Actual Work**.

- **Edits to total actual costs will be spread to the status date** – When this option is selected and a **Status Date** is set in the **Project Information** form, then Actual Costs are spread from the Actual Start to the Status Date. Actual Costs should not be entered in a time-phased form such as the Task or Resource Usage forms, but the results may be viewed in these forms. **Actual costs are always calculated by Project** must be unchecked for this option to available.

- **Default fixed cost accrual:** – This option sets the default accrual method for **Fixed Costs** (covered in **ASSIGNING RESOURCES AND COSTS TO TASKS** chapter). Fixed costs may be accrued at the **Start**, **End**, or **Prorated** over the duration of the task.

22.4   Proofing

This form is used to decide how the spell checker works and is similar to other Microsoft products and will not be covered in this book.

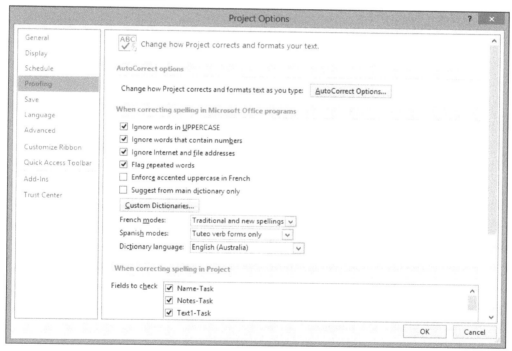

22.5   Save

This form specifies the file format, location and Auto save options.

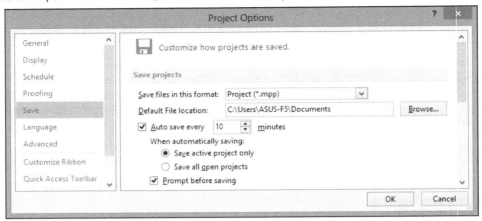

22.5.1   Save projects

- **Save files in this format:**

  There are several formats that your project may be saved in. The default file type may be defined here but the file type may be changed at the time a project file is saved:

  ➢ **Project (*.mpp)** – The normal format to save Microsoft Project 2010, 2013 and 2016 files.

  ➢ **Microsoft Project 2007 (*.mpp)** will save files in this format as a default.

  ➢ **Microsoft Project 2000 – 2003 (*.mpp)** will save files in this format as a default.

  ➢ **Template (*.mpt)** – Use this format to save projects that you want to use as project templates. Save them in the **User templates** directory that is specified below.

  ➢ **Microsoft Project 2007 Templates (*.mpt).).**

  ➢ **XML format (*.XML) – Extensive Markup Language** which is increasingly being used by other software packages like Primavera and Asta Powerproject to import and export data

  ➢ A number of other formats are listed but these are mainly for importing and exporting data not saving complete schedules that may then be opened and rescheduled.

  ➢ **ODBC Database** – Your project may be saved as an ODBC compliant database.

  Earlier Microsoft Project versions had some or all of the following options that are no longer available:

  ➢ **Microsoft Project 98 (*.mpp)** – The Microsoft Project 98 format which supported in some Microsoft Project versions released after Microsoft Project 98.

  ➢ **Microsoft Project Exchange (*.mpx)** – If you have Microsoft Project 98, you may save the file in **mpx** format for transfer to other older scheduling programs such as P3 and SureTrak.

  ➢ **Project Database (*.mpd)** – This is a database format, which allows multiple projects to be saved in one database. It may be used for exporting data to other programs.

  ➢ **Microsoft Access Database (*.mdb)** – A part or all of a project may be saved in Microsoft Access format and may be read and modified by Access.

- **Default File location:** – Used to set the default locations for your project data files. **NOTE:** If you are working in the same directory all the time it is well worthwhile setting this directory.

- **Auto save** – These options are self-explanatory.

22.5.2 Save templates

Project templates allow organizations to create standard project models containing default information applicable to the organization. In particular, a calendar with the local public holidays should be created along with Views with redefined headers and footers to suit your organization. Therefore to save time when you create a new project, you should generate your own templates to suit the different types of projects your organization undertakes.

Default user template location: – This is where you set the default locations for your project templates.

22.5.3 Cache

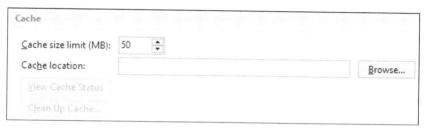

This allows users who are saving their files to a Project Server to speed up response time by saving a copy of the file locally. This option requires the Professional Version of the software.

22.6 Language

This form specifies the language preferences and will not be covered in this book.

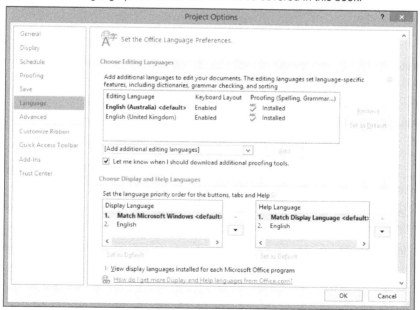

22.7 Advanced

This is where the advanced options may be edited:

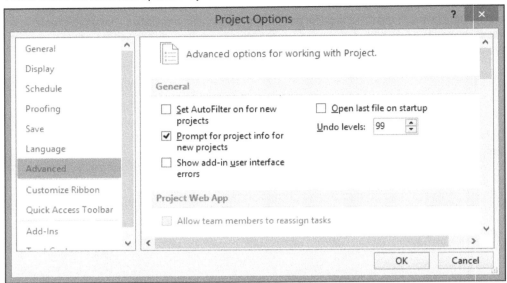

22.7.1 General

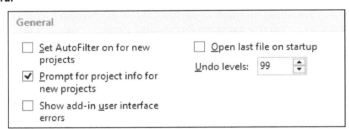

- **Set AutoFilter on for new projects** will display the down arrow by column headings for all new projects which in turn will allow an **AutoFilter** to be used immediately. This function may be turned on or off at any time by clicking on the **Display AutoFilter** button.

- **Prompt for project info for new projects** will open the **Project Information** form when a new project is created from the **Blank Project** template, which uses the **Global.mpt**. **NOTE:** This does not work with other templates.

- **Show add-in user interface errors**. An error will be displayed when this is turned on and add-in software attempts to control the Microsoft Project user interface.

- **Open last file on startup** will open the last file when Microsoft Project is opened which is useful if you only work on one project.

- **Undo levels** may be increased to 99 in Microsoft Project 2007. Earlier versions have one undo.

   This is a keystroke undo and when you have two files open it will undo changes made in both files. Undos are lost when a project file is saved.

22.7.2 Project Web App

Project Web App

☐ Allow team members to reassign tasks

Allows team member to reassign tasks is used with the Microsoft Project Web App as part of Microsoft Project Server. Please see chapter for an introduction to the Microsoft Project Server.

22.7.3 Planning Wizard

Planning Wizard

☐ Advice from Planning Wizard
    ☑ Advice about using Project
    ☑ Advice about scheduling
    ☑ Advice about errors

Planning Wizard – These options specify what advice the Planning Wizard will offer you and applies to all projects.

 You may find it is better to switch some of the options off once you know how to use the software as they often encourage users to make a change they should not make. Once you know how to use the software it wastes time closing these forms.

22.7.4 General Options

General options for this project: 🛢 OzBuild Bid Workshop 18 ▾

☑ Automatically add new resources and tasks
Default standard rate: $0.00/h
Default overtime rate: $0.00/h

These options may be applied to selected projects or all new projects.

- **Automatically add new resources and tasks** – This option, when switched off, confirms if a resource exists in the resource pool. If the resource does not exist, a form will ask for confirmation to add the Resource to the resource list.

- **Default standard rate:** is the default standard hourly rate assigned to new resources.

- **Default overtime rate**: is the default overtime rate assigned to new resources.

 These rates are not used when work is assigned against a task without a resource being assigned. This is different to the way Oracle Primavera P6 works.

22.7.5 Edit

Edit

☑ Allow cell drag and drop   ☑ Ask to update automatic links
☐ Move selection after enter   ☑ Edit directly in cell

- **Allow cell drag and drop** – Allows you to select a cell and drag the contents to another location.

 This needs to be switched on in order to be able to drag columns.

- **Move selection after enter** – After completing an entry in a cell, pressing the **Enter Key** will result in the cursor moving down a line to the cell below.

 Some users find it a nuisance when the cursor moves down every time they press the enter key. Uncheck this option If you wish the cursor to stay in the edited task.

- **Ask to update automatic links** – On opening a file which has dynamic links to other programs such as Excel, you will be asked if you want to refresh the links to the most recent data. To create a link, copy the selected data from another program. Then, in Microsoft Project, select the **Paste Link command** and the data will be placed in the location of the mouse, which may be in the bar chart area or a cell. See the **ADDING TASKS** chapter for more details.

- **Edit directly in cell** – Allows you to edit the data directly in the cell. Otherwise data has to be edited in the **Edit Bar** near the top of the screen.

22.7.6 Display

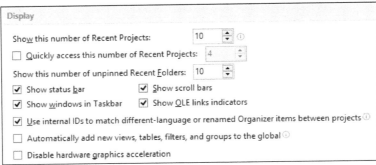

- **Show this number of recent documents:** allows up to 50 recently opened project files to be listed an increase from 9 in Microsoft Project 2007.

- **Quickly access this number of Recent Projects:** Allows the user to quickly access Recent Projects in the **File** tab.

- **Show this number of unpinned Recent Folders:** sets the number folders displayed when you select **File, Open, Computer**.

- **Show OLE links indicators:** This shows cells with links with the indicator links

- The other "**Show**" options allow each of the display options to be hidden or displayed.

- **Use internal IDs to match different-language or rename Organizer items between projects** is used in conjunction with Organizer when importing items from another project.

- **Automatically add new views, tables, filters, and groups to the global** was a new function to Microsoft Project 2010 which automatically adds any new views, tables, filters, and groups to the Global.mpt.

 Therefore as time progresses with the option **Automatically add new views, tables, filters, and groups to the global** activated, a project that is created from the **Blank Project** will have many Views, Tables and Filters from old projects that may be irrelevant to the current project and it is suggested this is turned off.

- **Disable Hardware Graphics Acceleration** is also called as **GPU rendering**. This is used to increase the program performance, but sometimes it can cause problems on old computers and some new computers causing blurred text/font problem and random cursor hanging problem.

22.7.7 Display Options for this Project

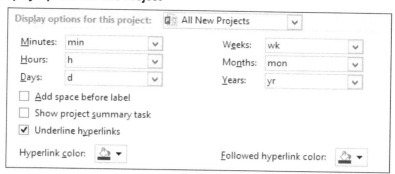

- **Minutes:**, **Hours:**, **Days:**, **Weeks:**, **Months:**, **Years:** – From the drop-down boxes, select your preferred designators for these units.

- **Add space before label** – Places a space between the value and the label; this makes the data column wider.

 It is **RECOMMENDED** changing "days" to "d" and "hr" to "h" to make the duration columns narrower and to uncheck **Add space before label** to make duration columns narrower.

- **Show project summary task** will display a virtual Task ID 0, which will summarize all project tasks. A Project Summary Task:

  ➢ May not be assigned resources,

  ➢ Is normally displayed as a gray bar

  ➢ The Task Name is linked to the **File**, **Info**, **Project Information**, **Advanced Properties** form **Summary** tab **Title:** field.

- **Hyperlink appearance in 'Project 1':**

  ➤ **Hyperlink color:** and **Followed hyperlink color:** – Select from the drop-down box the color you require for these data items.

  ➤ **Underline hyperlinks** – Select if you want these data items to be underlined.

22.7.8 Cross project linking options for this project:

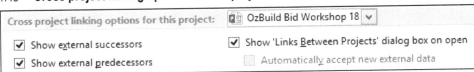

Cross project linking options for 'Selected Project' are options for displaying predecessor and successor task information from other projects. Inter-project relationships are not covered in this book.

 It is recommended that inter project links are made from milestone to milestone so they are more visible.

22.7.9 Earned Value options for this project:

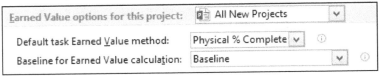

- The **Default task Earned Value method** (a new field in Microsoft Project 2002) may be set to use the task **% Complete** or the **Physical % Complete** to calculate the Earned Value.

 The **Physical % Complete** is useful when the Earned Value is not progressing in proportion to the duration and may be used for recording the actual progress of the deliverables of a task as opposed to % of the Work (recorded in % Work field) or % of Duration (recorded in the % Complete field).

- The **Baseline for Earned Value calculations:** allows the selection of any of the baselines for the calculation of the Earned Value.

 Strong corporate discipline and procedures are required to collect and enter Earned Value data at activity level.

It is recommended that Custom Fields and Custom Field formulae are used to calculate Earned Value in Microsoft Project as you will be able to set how the summary values are calculated. See the picture in para 21.8 Summary Tasks and Earned Value to see how Microsoft project adds up the Earned Value, observe the VAC Column.

22.7.10 Calculation options for this project:

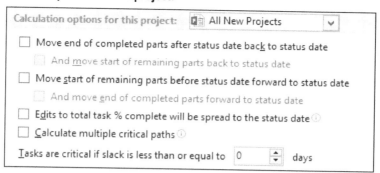

- New functions have been introduced in Microsoft Project 2002, which are intended to assist schedulers to place new tasks in a logical position with respect to the **Status Date**. The following four options were new to Microsoft Project 2002 and are covered in detail in the **Status Date Calculation Options** para 24.9.3.

  - ➤ **Move end of completed task parts after status date back to status date**
  - ➤ **And move start of remaining parts back to status date**
  - ➤ **Move start of remaining parts before status date forward to status date**
  - ➤ **And move end of completed parts forward to status date**

These functions do not work intuitively and the instructions in the Microsoft documentation do not describe these functions clearly. Review and understand these options before applying them.

- **Edits to total task % complete will be spread to the status date** – When this option is selected and a **Status Date** is set in the **Project Information** form, and Progress Lines are being viewed then the % Complete is spread from the Actual Start to the Status Date which may be viewed in the **Task** or **Resource Usage** forms.

- **Calculate multiple Critical Paths** – When checked, tasks without successors have their Late Dates set to equal their Early Dates, are calculated with zero Total Slack, are indicated critical in the Critical column and displayed as critical in the bar chart. (This is the same function as the Oracle Primavera software **Open Ends** options.)

  This function is useful if a project has multiple Areas or the Contract has several delivery dates and it is required to create a Critical Path for each Area or Delivery date.

  Calculate multiple Critical Paths – Unchecked, all float is calculated to end of the project.

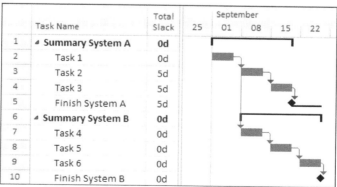

Calculate multiple Critical Paths – Checked, each network has its own Critical Path.

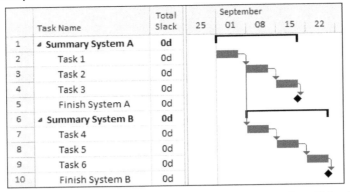

- **Tasks are critical if slack is less than or equal to ? days** – This option will flag tasks as being critical that have **Total Slack** (Total Float) less than or equal to the value entered in this form. Critical tasks may be displayed in the **Critical** column and on bars when the "critical bar" is displayed. This is useful for flagging "near critical" tasks and should be used when multiple calendars are used to identify the critical path.

22.8   Customize Ribbon

This is where commands may be added to and removed from the Ribbon. This works in the same way as other Microsoft products so will not be covered in detail:

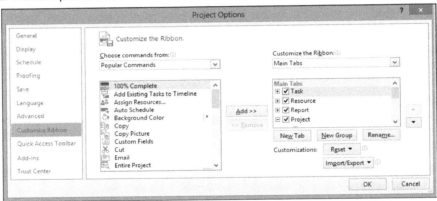

22.9   Quick Access Toolbar

This is where commands may be added to and removed from the Quick Access Toolbar.

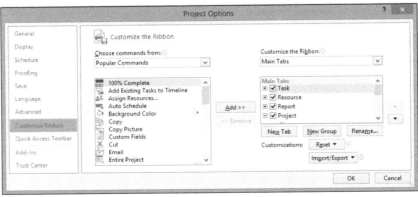

One of the more noticeable differences in Microsoft Project 2016 is that it now has put large spaces between the buttons in Quick Access Toolbar, making the Quick Access Toolbar less useful than in earlier versions. In the two pictures below you may see that there are many more Quick Access Toolbar buttons in the Microsoft Project 2013 screenshot than the Microsoft Project 2016 screenshot:

Microsoft Project 2016

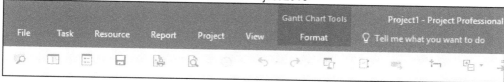

Microsoft Project 2013

 The introduction of the **Ribbon Toolbar** has increased the number of keystrokes required to access commands compared to the traditional toolbars from earlier versions of Microsoft Project. Users may find it simpler to hide the Ribbon using the **Minimize the Ribbon** command, move the **Quick Access Toolbar** below the **Ribbon** and add all the frequently used commands to the Quick Access Toolbar, as they are used, by right-clicking on them. The order may be revised using the **File**, **Options**, **Quick Access Toolbar** tab.

22.10 Add-Ins

An add-in is installed functionality that adds custom commands and new features to Microsoft Office system programs. Add-ins can be for various types of new or updated features that increase your productivity.

This form is where Add-ins, additional pieces of software that operate with Microsoft Project, may be viewed and managed:

22.11 Trust Center

The **Trust Center** allows the control of privacy and macro security.

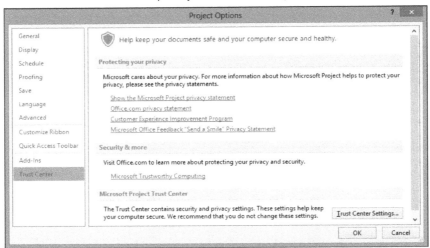

The **Trust Center** may be opened by:

- Clicking on the **Trust Center Settings...** button, or

- From the **Developer**, **Code** group, ⚠ **Macro Security** button.

22.11.1 Macro Settings

Macro Security was a new feature in Microsoft Project 2003 and these options are set to nominate which macros may or may not be used. This is aimed at preventing macros with a virus being activated.

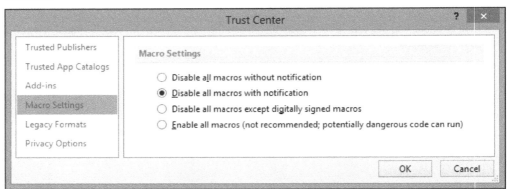

22.11.2 Legacy Formats.

These options, new to Microsoft Project 2007, control the notification and opening of legacy file formats such as Microsoft Project 98 format and Microsoft Project 2000 – 2003 mpp files.

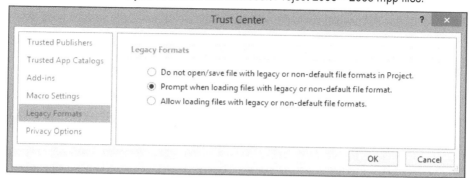

 The option in the picture above is recommended as you will not be prevented from opening a file created in an earlier version but will be informed that it is a **Legacy Format**.

It is recommended that you do not operate with files saved in a Legacy Format as this has been associated with file corruption. If a person requires a file in a Legacy Version then you should save the file with a new name indicating it is a Legacy Format, close it and reopen your original file. Avoid going backwards and forwards in versions.

22.11.3 Privacy Options

Privacy Options

The **Privacy Options** section allows you to determine what access Microsoft has to your machine.

Document-specific settings

The properties of Author, Manager, Company and Last Saved By are removed from a file when it is saved. It is recommended that this not be checked if you are taking your page setup header and footer from the **Project Information** form.

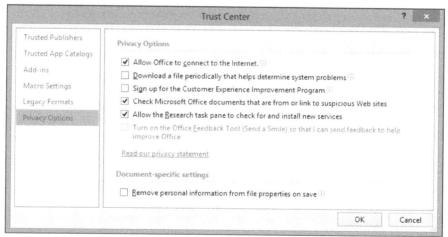

 People with projects that have many views often populate the headers and footers with data from the **Project Information** form. This allows all headers and footers to be changed from one place. In this situation you do not want to **Remove personal information from file properties on file save** checked.

22.12 Authors Recommended Setup for New Projects

Many Microsoft Project default options are set for advanced scheduling and should be avoided by inexperienced users.

The screen dumps in this chapter set to the options that the author would suggest are good defaults for inexperienced and casual users.

The correct setup of options is critical for minimizing the problems that users often experience and time should be spent understanding these options.

22.13 Commands Removed From the Microsoft Project 2007 Options, View Form

The following Outline options have been moved out of the Options form and are now only available as commands which may be added to the **Quick Access Toolbar** or **Ribbon**.

Indent name
- Indents the title for each Outline level to the right or removes the indent. It is useful to remove the indent, as displayed below, when there are a large number of outline levels and small manual indents may be created with a space at the start of each Task Name. This is also a useful function when the summary tasks have been hidden to right align all the tasks descriptions.

- The command ✓ **Indent Name** may be added to the **Quick Access Toolbar** or **Ribbon**.

Show outline number
- Shows the Outline Number 1, 1.1, 1.1.1, etc.; see picture below.

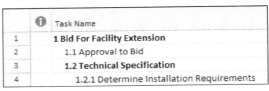

- The command ✓ **Outline Number** may be added to the **Quick Access Toolbar** or **Ribbon**.

Show outline symbol
- Displays the ◢ sign by the Task Name, see picture below:

- The command ✓ **Outline Symbol** may be added to the **Quick Access Toolbar** or **Ribbon**.

Show summary tasks
- Uncheck this to hide the summary tasks in the active view.

---

23 MICROSOFT PROJECT SERVER NEW

This chapter was written by Martin Vaughan, of Core Consulting Group in Melbourne Australia and is reproduced in this book with their permission. Core Consulting Group are Project Management Office (PMO) Specialists who provide Services and Solutions across Portfolio Management, Program Management and Project Management Offices and specialize in EPM implementations. They may be contacted through their web site http://coreconsulting.com.au.

23.1 Understanding Microsoft Project Server

Microsoft Project Server is an **Enterprise Project Management** (EPM) tool, used to help manage a Portfolio of Projects and Programs. It combines a centralized database, a web based application, SharePoint lists used for collaboration plus additional functionality over and above Microsoft Project Professional.

From the Scheduler/Project Manager perspective, they effectively use three applications:

1. Web based **Project Web App** (PWA) for most of the Portfolio Management functions and to manage project level information (e.g. Status Reporting). This tool is also the key interface if Workflow, Timesheeting or Resource Management are deployed.
2. Web based SharePoint "Project Site" for collaboration, document management and lists (e.g. Risks, Issues, Change Requests).
3. Microsoft Project Professional for planning/tracking of the schedule.

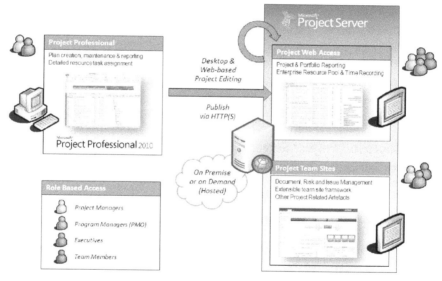

The EPM system provides an extensive amount of additional functionality to help manage the Portfolio, including:

- Integration to project methodology including workflow (automation of decisions),
- "Pipeline" or Demand management (selection and prioritization of upcoming projects),
- Resource Management across the portfolio,
- Timesheeting (recording of actual hours),
- Finance system Integration (which usually involves customization),
- Extensive reporting and dashboard capability,
- Centralized accessible data, single source of truth.

Benefits from an EPM system are hard to quantify, but will typically be a combination of:

- Improved visibility, and hence better decision making. This may even extend to rationalization of projects and improvements in investment efficiency,
- Efficiency gains through standardization, removal of duplication of reporting and through workflow automation,
- Improved transparency through record of decisions.

23.2 Changes for Schedulers and Project Managers

EPM provides an environment where most project information can be kept in a single location in a consistent manner. The main change from stand-alone Microsoft Project Professional for Schedulers/Project Managers is that they will need to plan and control in a defined manner, not necessarily how they have "always done it".

One of the key success criteria for adopting EPM is the definition of a standard for planning, scheduling, tracking and reporting. Standards will be defined by each organization according to their needs but typically cover:

- The "rules" in **Options** are centrally set via the **Global**, which also contains **Calendars**, **Views**, **Tables**, **Filters**, **Groups** etc. are set for all users,
- Structure of schedules,
- Level of detail expected,
- Summary/Task/Milestone naming conventions,
- Custom field use,
- Scheduling approach – e.g. using logic & durations or just enter dates,
- Resourcing approach – e.g. use "bucket tasks" or resource the detail,
- Tracking approach – e.g. using Actual Dates and Remaining duration or % complete,
- More advanced considerations such as Baselines, Inter project dependencies, rolled up reporting.

23.3 How Microsoft Project Server Operates

From a "hands on" perspective, there are some key aspects of a typical EPM implementation worthy of further discussion.

23.3.1 Server connection

Users need to define the location of their EPM in Microsoft Project Professional. This would have already been done as part of the rollout of EPM ideally. Under the **File**, **Info** menu there is a **Manage Accounts** section where a URL must be defined. Depending on the option chosen, as Microsoft Project Professional starts in the future, a connection will be made. Importantly, Microsoft Project Professional will then adopt the "Enterprise Global" Options not your PC's Options. Using the **File**, **Open** menu, the list of projects will then also enable browsing of the projects inside of EPM. This picture is from Microsoft Project 2016 but the process is the same with Microsoft Project 2013.

23.3.2 Security

Whilst EPM promotes collaboration, there is also a defined security model which restricts who can see what, and who can do what. Typically the model is as follows:

- Team members can see your schedule (but not edit it) and all other aspects of the project as well as raise Issues, Risks etc. They typically cannot see other projects they are not a team member of.

- Project Managers and/or nominated schedulers/controllers can edit schedule and other data. They typically can see other projects within their business unit.

- Program Managers can typically edit projects within their program.

- The PMO are typically "super users" who can see and edit all.

- Line Managers or Resource Managers have a different view (by resource) and may be able to edit assignment information depending on the EPM configuration, within their business unit.

- Senior Managers can typically see all but not edit.

There is often a security model over laid which hides visibility of sensitive projects and possibly projects in other business units. The model changes depending on the organization's EPM design and configuration, the above is indicative.

23.3.3 Check out/Check in

People used to Document Management systems would understand the nature of checking in and checking out a project. Effectively the "owner" of the data can Check Out a project by opening it. They cannot Check Out or open a project if someone else already has it open. This is not restricted to the schedule; another user could open or Check Out the project as they modify metadata or other information. While Checked Out, users will not be able to open the schedule for editing. This also works vice versa so users need to ensure they Check In a project as soon as they are finished editing information. While Checked Out, the project remains Read only and people can only see the previous version. When using PWA or Microsoft Project Professional, when you close a project, EPM should prompt you to Check it back in.

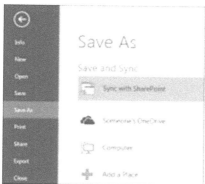

23.3.4 Publishing

There is a not so obvious function in EPM called **Publish**. Using a similar analogy to a magazine or newspaper, others cannot see your project data or use it in reports until you Publish. This effectively lets everyone know you are happy with the schedule and other data. Just because a schedule is Checked In, it doesn't mean it is Published. There is a menu command on the **File** menu for this, users click **Publish** before they Check In and close, otherwise the previously published version of the schedule will be used in reports. This is a good function which provides Project Managers with control over who sees what data relating to their project. But it is easy to forget.

23.3.5 Centralized Global

Refer to para 25.13 for more information about Organizer and the Global. Normally, when a user opens Microsoft Project on their PC, the local Global also opens. This is where default Microsoft Project Views, Tables, Filters, Groups, Calendars, Macros etc. are normally kept. Users can then create their own Views, Tables etc. and either keep them in a local Microsoft Project file, or put copies in Global for future use across any project on their PC.

With EPM, an Enterprise Global is used, which means a set number of Views, Tables etc. are centrally controlled. Changes made during a session will therefore revert back once Microsoft Project Professional is closed. For this reason users should ensure that any customizations to Views, Tables etc. are done on separately created and named versions of those Views, Tables etc. For example, rather than re-formatting "Gantt Chart", create a new view such as "My Gantt Chart" and format that.

The other impact of an Enterprise Global has is on the "rules" or Options which affect the behavior of the tool. People who change Options, for example they may want to turn off "Effort Driven", will be governed by the Enterprise Global. Users need to familiarize themselves with the Enterprise Global Options and if they have concerns raise them with the PMO and/or EPM Administrator.

23.3.6 Centralized Resource Pool

One obvious benefit of EPM is that it centralizes data. From a resource perspective the list of resources, which users typically see and edit via the **Resource Sheet** is managed centrally. There are many reasons for this including naming conventions, integration to Active Directory (security/passwords), licensing and standardization of cost rates. It also enables Resource Management, Timesheeting and collaboration functions. Users can also have "local resources", where they do not exist in the Enterprise pool.

From the **Resource** tab, select **Add Resources** then **Build Team from Enterprise**, this will allow users to select which resources they would like to use. There is also an option via the **Assign Resources** form in Microsoft Project Professional which does a similar function. Note that the example list shown below contains both "real" resources as well as "generic" resources. This enables resource planning as well as concepts such as "Requested" and "Committed". The PWA showing the resources looks like the following:

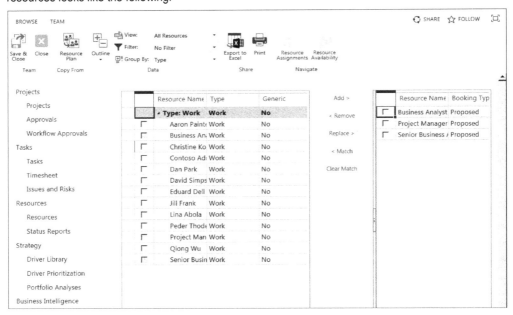

23.3.7 Modelling of resource effort and cost

While chapters 17to 0 of this manual covered extensively the resourcing and costing functions, when using EPM users will need to give consideration to the following:

- For Resource Management purposes, all effort must be represented somewhere in a schedule (for the Enterprise that means even Business as Usual / BAU activity and minor initiatives). There are many and varied approaches to doing this.

- It is effort which must be modelled, not duration, so assignment of a resource at a percentage unit is necessary. This means users must be familiar with the function of the tool and the data entry steps so as not to affect task durations unwittingly.

- There are a number of ways to represent cost in a schedule. We find the best way is through the use of a cost type of resource rather than the Fixed cost column. EPM 2010 had issues in the web environment with this.

Shown below is a single schedule highlighting how best to represent fixed costs and resource costs. Note the use of "Bucket tasks" to model resource effort. A bucket task is a long task, spanning across detailed tasks/milestones for the purpose of modelling resource effort and hence cost. This approach would be appropriate for Business/IT projects plus some construction/infrastructure projects and would be particularly useful for Timesheeting. It may not suit all projects, particularly those scheduled to the nearest hour. If the schedule is to be used for Resource Management and/or cost estimating, this bucket task approach is not only simpler but far more accurate we think.

Note the use of Milestones for cost, in this case the hypothetical project has payment by milestone. It would also be possible to model monthly progress payments this way. In all cases the milestone dates would need to align to when invoices are expected. Also note that individual tasks/milestones can be aligned to Finance system general ledger codes. Finally, line 51 is a good example of how Contingency, both time and cost can be represented within each phase.

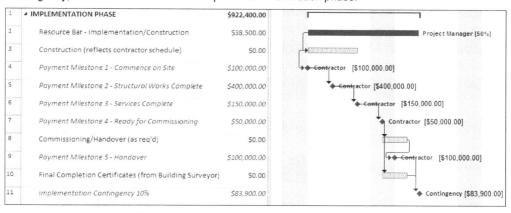

The key to using Microsoft Project and EPM for Cost tracking, we believe, is to only use it for the Estimate to Complete (ETC) or Remaining Cost. Using SharePoint functions in EPM, we can then collect Actual Costs and combine them with the ETC to provide full cost reporting.

This is complex and beyond the scope of these notes. For further information contact Core Consulting Group on +61 3 9654 0561 or info@coreconsulting.com.au.

23.3.8 Updating Schedules Properly

As covered in para 25.13 there are many ways to update a schedule. When using EPM's Resource Management, Timesheeting and/or Costing functions, there is a need to reschedule future activities into the future with respect to the Status date. This makes consistent and thorough updating of schedules a priority. These techniques may be new to some Schedulers and Project Managers who are used to tracking using percent complete however.

- Schedule updated % complete incorrectly with incomplete work in the past and no Baseline:

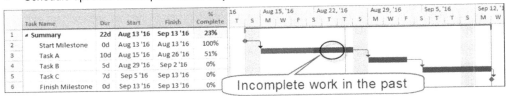

- Schedule updated properly with a baseline and all incomplete work in the future and complete work in the past:

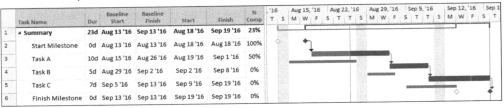

23.3.9 Project Inter-Dependencies

Modelling of inter project dependencies is always challenging, there are a number of ways to implement. EPM resolves the file naming and location problems normally associated with inter project dependencies, through the very nature of the centralized data base.

There are various solution options available. Microsoft promotes the use of PWA and functionality embedded within Deliverables lists. Other organizations work with predecessors directly from other schedule files. Some use customized solutions to pass through just the latest forecast dates to inform recipient Project Managers but empower recipient Project Managers to accept or resolve date shifts.

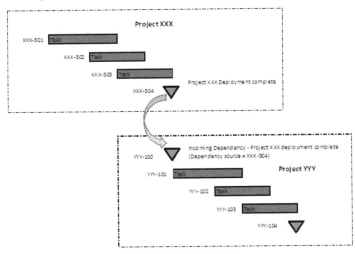

23.4   New Functions for Project Managers

It is beyond these notes to go into too much detail about additional EPM functions. A brief summary of new functions for mainly the Project Manager, or those who assist the Project Manager, are as follows:

- **Issues**, **Risks**, **Changes** – There are SharePoint lists and forms which are configured to keep track of things. The most important ones for Project Managers are likely to be the Issues, Risks and Change logs. Additional logs can be easily set up by the PMO, e.g. to track Variations or Lessons Learned. These logs are reasonably intuitive to use but Project Managers may need guidance as to when to assign various values on the drop down menus.

- **Document Management** – SharePoint has extensive functionality to assist with managing project documents.

- **Finance integration** – Primarily for Actual cost but occasionally for funding and budgets this is a complex area. Done correctly, Project Managers will see the actual costs being booked at an aggregated monthly level by GL code or at the project level. If they wish to drill down they will need to use their finance system to look at the detail. The key aspects here are that finance data is typically only accurate at month end and that the finance system is the "single source of truth" for some data.

- **Timesheeting** – Depending on the deployment model, Project Managers may need to approve time booked to their project by project team members. There is additional functionality available in PWA for this. Project Managers will also need to fill out their own timesheet.

- **Resource Management** – Depending on the deployment model, Project Managers may need to negotiate with Line Managers for resource commitment. Using the resource graphs in PWA, workload across multiple projects can be assessed by the Line Managers and others. With the careful use of generic resources a demand model can be created and resources managed by skill set.

- **KPIs and Commentary** – Key Performance Indicators (KPIs) are often included in EPM deployments, calculated by the tool or selected by the Project Manager. Often these KPIs are "remembered" from previous months so trends can be clearly shown on reports. More often than not EPM Status reporting includes commentary, at the project level, against KPIs and against individual schedule items. This commentary will need to be added by the Project Manager or their assistants. It really depends on what the organization deploys as to the extent of KPIs and commentary.

- **Project Status Reports** – Again each deployment of EPM is different, but typically Project Status reports are compiled and verified by the Project Manager before they are Published. Reports can bring information in from schedules, from the logs and from PWA forms. This occurs typically monthly, at the end of the month to align with finance end of month. This means schedule status needs to be aligned at the same date and the schedules published too. Portfolio Summary reports will also be published once all project status reports have been completed.

23.5   Tips for Implementers

It is beyond the scope of these notes to go into too much detail about deploying EPM. The key point is that is a journey, not a big bang overnight change. The most successful EPM implementations we have seen have involved the organization clearly identifying what they do and don't need in terms of functionality before kicking off the requirements gathering. We call this step the "Concept of Operations". Our key tips are:

- Keep it simple, like most things in life it is also a key theme with an EPM deployment. It is so easy to be overwhelmed with complexity, especially with finance integration. Spare a thought for the end user, it will be quite a change.

- This leads us to the next recommendation, focus on Change Management and user adoption. The best configured tool will provide little value if people don't know how or won't use it. Communications, training and alignment to standards are critical, as is cleansing and preparation of data.

- Deploy the tool in stages, initially focusing on gaining a "single source of truth". Once everyone is using it, the organization can then focus on process change and efficiency improvements. Rolling out a change in tool at the same time as a change in process increases risk and complexity enormously.

If you are thinking about, or need help with an EPM deployment, please contact Core Consulting Group on +61 3 9654 0561 or info@coreconsulting.com.au.

23.6   Accessing Projects from a Microsoft Project Server

There are two methods of opening a project from a Microsoft Project Server:

- From a desktop version of Microsoft Project and Check Out the project, which is basically downloading the project onto your computer. Make the changes and then check the project file back into the server, which is effectively uploading the project file back to the server, or

- Use Internet Explorer to open a project on the server using the Web application.

Irrespective of which process you use you will require a Server Name, Server URL, User Name and Password that is set up by an administrator, an example of how this would look is below:

- Server Name   Eastwood Harris

- URL: http://epm.eh.com/test

- User Name: eh\paulharris

- Password: password

This section will develop this topic further and demonstrate some methods of open files from a server.

23.6.1 Opening a Schedule from Microsoft Project

To open a project from a Microsoft Project Server:

- Select **File**, **Open** the **Open** form has a new option of **EPM** that allows you to **Browse** the server and open project from the server.

- You may click on the **Show me the list of all projects** which then opens a second form and you may select a project to open:

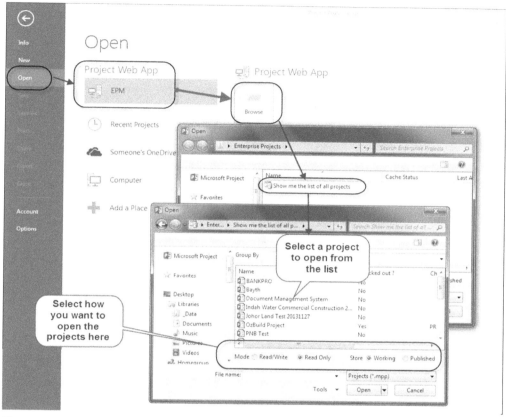

The project will be opened in the normal way ready for you to work on it.

23.6.2 Creating, Saving and Closing a Schedule

A project may be created with Microsoft Project Professional in the usual way by using the **File**, **New** command. Your Global Options may be changed by the Project Server Options on save and your schedule may calculate differently when it is reopened.

The command **File, Save As** will give the options of where to save to the schedule:

- Select **File**, **Save As** to open Save to Project Server the form:

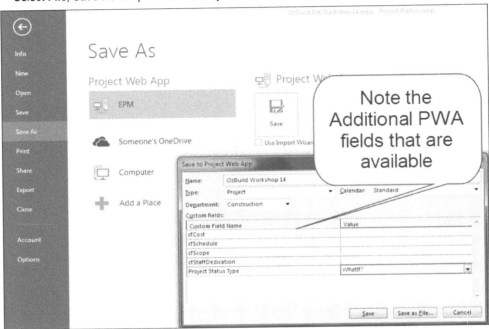

- As a minimum enter the project name and click on the ⎡ Save ⎤ button to save to the server, or

- Click on the ⎡ Save as File... ⎤ button to save to your computer and select the appropriate option.

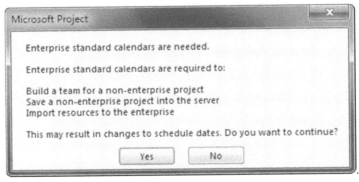

23.7 Accessing Project Server Schedules through the Web App

To access a schedule on the **Web App** you will first need to log on to the server and then open the project from the server.

To log on to the server:

- Open **Internet Explorer**, other browsers will not operate,

- Enter the URL of the sever and you will be presented with the **Windows Security** login screen, enter your **User Name** and **Password** and

- Click [OK] to continue and you will be taken to the Project Server **Home** page. An example of a "Home Page" is below but depending on how your home page has been customized by your organization it may look completely different to the picture below:

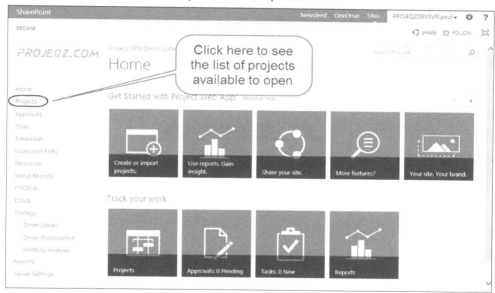

Access to a Project Server for these server screen shots was provided by Vik Subramaniam of PROJEQZ.COM of Suite D-39-2, Zenith Corporate Park, No. 1 Jalan SS7/26, Kelana Jaya, 47301 Petaling Jaya, Selangor DE, MALAYSIA. Tel: +60378052487, emai: vik@projeqz.com. If you have any Microsoft Project Server consulting or implementation requirements please contact PROJEQZ.COM.

- Select **Project Center** to see a list of projects you have access to, and

- Select the project you wish to open by clicking in the left hand column:

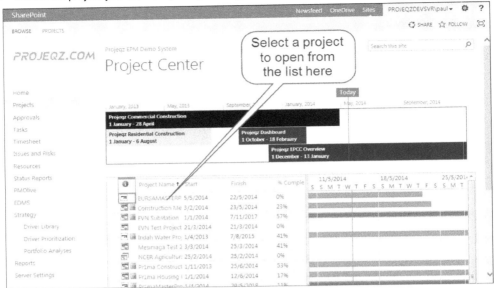

- Select how you wish to open the project from the drop down box under the **Open** button or click on the icon to the left of the project name:

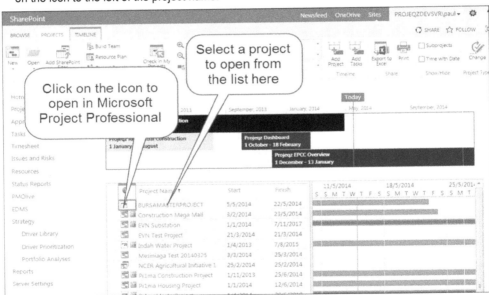

- If you select **In Project Web App for Editing** then you will be able to edit your project in the **Web App**.

In the Web App you will find less functionality than using the desktop software but allows the schedule to be edited without a desktop copy of Microsoft project Professional.

24 MORE ADVANCED SCHEDULING

The following advanced scheduling options should be considered once a user has some experience in the use of the software.

24.1 Working with Hourly Calendars

Hourly calendars are used for short duration projects such as medical operations, events and plant shutdowns. Hourly calendars add more complexity to a project and this section covers these issues.

24.1.1 Adjusting Calendar Default Working Hours

To adjust the standard working hours of a calendar:

- Select **Project**, **Properties** group, **Change Working Time** to open the **Change Working Time** form:

- Select the calendar to be edited from the **For calendar:** drop-down list,

- Click on the **Work Weeks** tab,

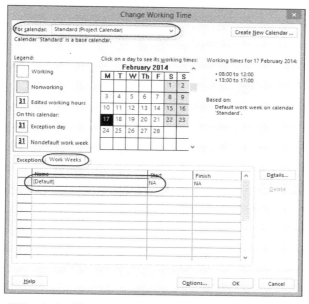

- Select the **[Default]** line in the **Work Weeks** tab,

- Click on the **Details...** button, or double-click on the **Name** to open the **Details for 'Calendar'** form,

- Days that are to be set the same working hours may be **Ctrl-Clicked** or **Dragged** to select them,

  The working hours for select day/s may be edited in the **From** and **To** cells on the right of the form.

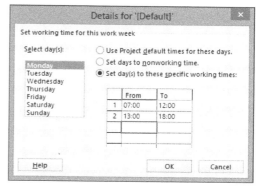

 To create a 7-day per week calendar you will need to set up Saturday and Sunday working times the same as Monday to Friday.

24.1.2 Creating Calendar Periods with Alternate Working Hours

This **Work Weeks** tab may be used to create a period of one or more days where the working hours are different from the default working hours. Longer working hours may be required during a system upgrade or shutdown period. To create a period with different working hours from the default:

- Select the calendar that is required to have a special work period. The example below uses the Standard calendar,

- Select the **Work Week** tab,

- Select the first blank line in the **Work Weeks** tab and type in the description for the period. The example below has a unique set of working hours for a February Shutdown,

- Enter the Start and Finish dates of the period in the **Work Weeks** tab,

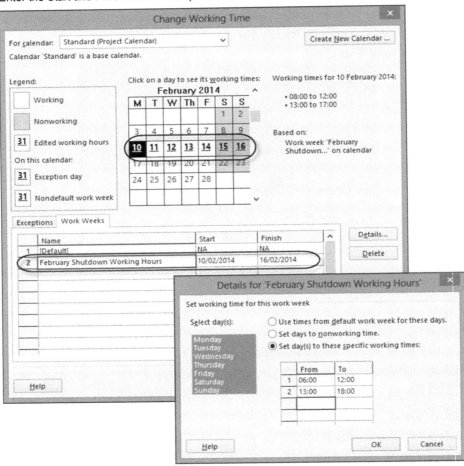

- Click on the ☐ Details... ☐ tab to open the **Calendar**, **Details for 'Calendar'** form,

- Select the days to be edited by dragging with the mouse pointer or **Ctrl-clicking**,

- Click on the appropriate radio button and edit the hours as required:

 When this option is used and the hours per day are not the same for each day then the durations displayed in days may be misleading. Options to manage this issue are discussed later in this chapter.

24.1.3 Working with Calendars with Different Hours per Day

There are some workable options to ensure that the durations in days are calculated and/or displayed correctly:

- All the calendars used on a project schedule should have the same number of hours per day for each day. This value is entered in the **Hours per day:** field in the Options form below. Then all durations in days will calculate correctly.

- When there is a requirement to use a different number of hours per day (in either the same calendar or in different calendars) then all durations should only be displayed in hours and the Task Calendar displayed, as per the picture below. Two examples of how this may occur are when:

  ➤ There is an 8-hour working day Monday to Friday and a 4-hour working day on Saturday, or

  ➤ Different calendars are required on a project, for example when some tasks are assigned an 8-hour per day calendar and some a 24-hour per day calendar.

  The **Duration is entered in:** field in the **File**, **Options**, **Schedule** tab, **Scheduling options for this project** section should be set to **Hours**.

- A Customized Field may be used to calculate and display the correct duration by using a formula. The example below uses the **Duration1** field to calculate and display the correct duration in days of a 24-hour per day calendar using a formula.

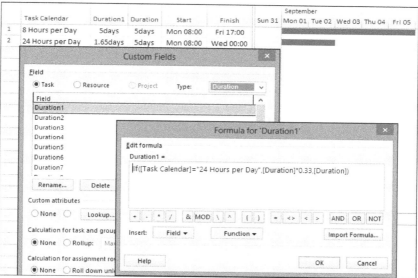

To create the formula displayed above:

- Select **Project**, **Properties**, group, **Custom Fields** to open the **Custom Fields** form,
- Select **Duration 1** from the **Type** drop-down box in the top right-hand side of the form,
- Click on the ☐ Formula... ☐ button to open the **Formula** form and enter the formula as required.
- The formula in the picture above will work for a schedule that has two calendars in use.
- The formula above is

      IIf([Task Calendar]="24 Hours per Day",[Duration]*0.33,[Duration])

24.1.4 Understanding Default Start and Default End Time

The **Default start time:** and **Default end time:** are the times that the software uses when a date is entered and a time is not entered. These times should be aligned to the **Project** calendar. They are used in Microsoft Project when:

- Constraints are assigned to tasks when a date is typed into a Start or Finish field, and
- Actual Start or Actual Finish dates are assigned.

These times are set in the **Options**, **Calendar** form which may be accessed by:

- Clicking the [Options...] button from the **Change working time** form, or
- Selecting the **File**, **Options**, **Schedule** tab, **Calendar options for this project:** section:

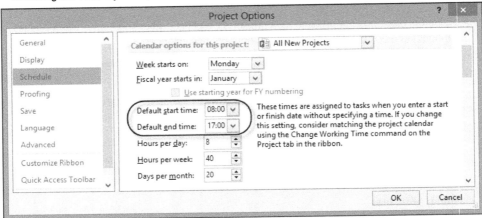

If these times are not aligned then tasks may be displayed one day longer than the duration as per the picture below, where the calendar start time is 8:00am and the Default start time is 9:00am and a 3 -day task spans 4 days:

Duration	Start	Finish	Monday	Tuesday	Wednesday	Thursday
3d	Mon 09:00	Thu 09:00				

24.2 Managing Calendars

24.2.1 Deleting a Calendar

To delete a calendar:

- Select **Developer**, **Manage**, **Organizer**, and select the **Calendars** tab.

- Highlight the calendar you want to delete and click on the [Delete...] button.

24.2.2 Copying Calendars between Projects

To copy a calendar between projects:

- Open both projects,

- Select **Developer**, **Manage**, **Organizer**, and select the **Calendars** tab,

- From the drop-down boxes at the bottom of the tab under **Calendars available in:** select the projects that you want to copy to and from, and

- Select the calendar you want to copy and click on the [Copy >>] button.

24.2.3 Renaming a Calendar

To rename a calendar:

- Select **Developer**, **Manage**, **Organizer**, and select the **Calendars** tab.

- Highlight the calendar you want to rename and click on the [Rename...] button to open the **Rename** form and type in the new name.

24.2.4 Copying a Base Calendar to Global.mpt for Use in Future Projects

Global.mpt is the default project template and all new projects created from a **Blank Project** copy their default settings from the Global.mpt.

Once you have set up a calendar with all the holidays for your location or business, it is suggested that you consider copying it to the Global.mpt to create your own personal template.

- Select **Developer**, **Manage**, **Organizer**, and select the **Calendars** tab.

- Select the **Standard** calendar from your project and click on the [<< Copy] button.

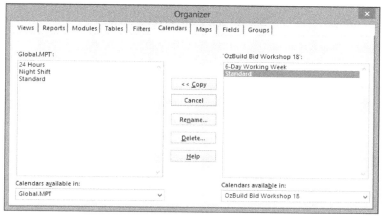

  If more than one person needs to use a tailored calendar then a template should be created and made available to all users.

24.2.5   Printing the Calendar

It is always useful to be able to print out the calendar for people to review the working hours and nonworking periods.

You may display the Calendar View by selecting **View**, **Task Views** group, **Calendar** which is discussed in the **TABLES AND GROUPING TASKS** chapter:

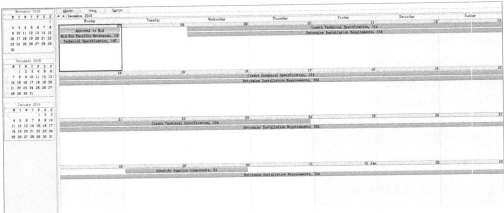

 If you open a Microsoft Project 2007 file and this Calendar View is not displayed, select **View**, **Task Views** group, **Rest to default** will display this calendar.

The **Calendar** view may hide some activities when there are many activities on one day. These will be printed out as text on a separate sheet, but you need to be careful when using the **Calendar** view.

It is useful to print the report to a pdf format so it may be saved and emailed.

24.2.6   Display Times in 24-Hour Clock

Should you wish to display your times in 24-hour format, e.g. 17:00hrs instead of 5:00pm then:

- Select from the windows menu **Control Panel**, **Region** tab, or similar commands if you are not using Windows 8.
- Select **Formats**, select the **Short Time & Long Time** tab and
- Change the time format from **h:mm:ss tt** to **HH:mm:ss** or **H:mm:ss**.

 This option will affect other programs that use the system settings but makes it simpler to edit times.

24.3 Adding Tasks

24.3.1 Task Scheduling Mode - Manually Scheduled or Auto Scheduled

Microsoft has introduced this new task attribute titled **Task Scheduling Mode** in Microsoft Project 2010. In earlier versions of Microsoft Project every task was set to the equivalent of the new **Auto Scheduled** setting in Microsoft Project 2010. In this mode all tasks' Start and Finish dates are calculated with respect to the Project Start (or Finish) Date, their Durations, Relationships, Constraints, Calendars and Options.

When a task is set to **Manually Scheduled** it does not need a predecessor or constraint to hold the Start Date and therefore will not move when the project is scheduled. When a task is **Manually Scheduled** it takes on the following attributes:

- It is clear from the default bar formatting and **Task Scheduling Mode** which tasks are **Manually Scheduled**.

- Tasks may have text entered as the Start Date, Duration and Finish Date, see Task A below.

- As more information is known they may be assigned Dates, Durations and Resources, see Task B below.

- As the tasks are finalized they may be converted to **Auto Scheduled**, see Task C below. This task is acknowledging the Project Start date of 12 July.

When two **Manually Scheduled** tasks are linked they acknowledge the relationship when:

- The **File**, **Options**, **Schedule** tab, **Scheduling options for this project:** section, **Update Manually Scheduled tasks when editing links** is checked, and

- A relationship is added or edited.

- In the pictures below:

  ➤ Tasks 9 has been dragged back in time and the Finish-to-Start relationship is not being acknowledged and

  ➤ The **Task Inspector**, which may be opened by selecting **Task**, **Tasks** group, **Inspect**, has highlighted the Finish Date of Task 9 and the outline of Task 9 is now dotted. This dotted outline will only display when the **File**, **Options**, **Schedule**, **Schedule Alerts Options**, **Show task schedule warnings** is clicked on:

Before change:

After Change

Summary and detail tasks may have different **Task Mode** settings:

- Summary tasks may be made **Manually Scheduled** and detail tasks made **Auto Scheduled**, as per Summary 1 in the picture below.

- The lower bar formatting and Finish date underlining of the Summary 1 is the **Task Inspector** function warning that there is conflict as the summary task duration does not match the detail task durations or dates.

- Summary tasks may be made **Auto Scheduled** and detail tasks made **Manually Scheduled**, as per Summary 2 in the picture below. In this situation the summary task adopts the duration of the detail tasks.

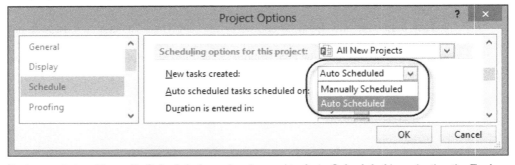

The default setting for new tasks is set in the **File**, **Options**, **Schedule** tab, **Scheduling options for this project:** section, **New tasks created:**

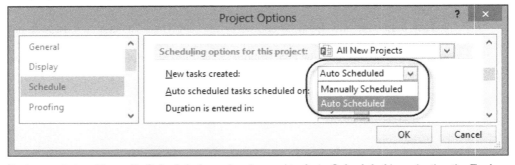

Tasks created as **Manually Scheduled** may be changed to **Auto Scheduled** by selecting the **Task Scheduling Mode** by:

- Displaying the **Task Mode** column,

- Clicking on the **Task**, **Task** group, ⚲ **Manually Schedule** or ⇥ **Auto Schedule** button.

- Right-clicking on the task text or bar and selecting the required option from the menu.

When a task is changed from **Manually Scheduled** to **Auto Scheduled** the **File**, **Options**, **Schedule** tab, **Scheduling Options for this project:** section, **Keep Tasks on nearest working day when changing to Automatically Scheduled mode** will:

- When checked it will set a **Start No Earlier** constraint on the task and it will not move forward in time even if it has no predecessors, or

- When unchecked the task will not have a constraint set and will move forward to the **Project Start** date when it has no predecessors.

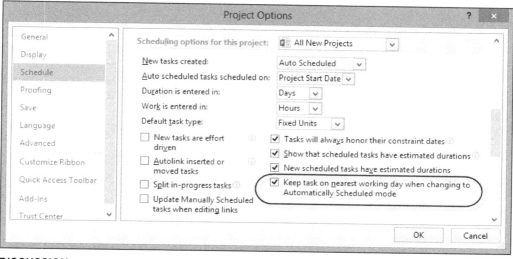

DISCUSSION:

- **Manually Scheduled** tasks would not be considered by many professional schedulers as good practice, as these tasks would not create a Critical Path schedule or meet most contract requirements.

- It could be a useful function during the initial development of a schedule but ultimately it would be considered good practice to have all tasks **Auto Scheduled** before a schedule is submitted for approval or to a customer as part of a submission.

 The **Auto Scheduled** option may be turned off so the schedule does not recalculate every time a change is made to the schedule by selecting **File**, **Options**, **Schedule** tab, **Calculation** section **Calculate project after each edit:**. To recalculate the schedule the user presses the **F9** key. This is not the same as setting tasks to **Manually Scheduled** and all tasks set as **Auto Scheduled** will be recalculated and **Manually Scheduled** tasks will not be recalculated when the **F9** key is pressed.

24.3.2 Copying Tasks from Other Programs

Task data may be copied to and from, or updated from other programs such as Excel, by copying and pasting. The columns and rows in your spreadsheet will need to be formatted in the same way as in your schedule before they may be pasted into your schedule.

Microsoft Project 2010 introduced new functions allowing:

- Copy and Paste to other products to keep the column header, formatting and indenting.

- Copy and Paste from other products to allow the option of keeping the source formatting:

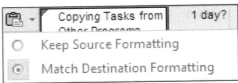

It is often best in Microsoft Project to make all Task Names unique, so for example when you have a building with many floors and trades each Task Name should include the trade and floor. This makes it easier to understand the schedule when a filter has been applied and to find predecessors and successors in a large schedule. These descriptions may be created in a spreadsheet by using the **Concatenate** function, the picture below demonstrates how. Text may be added by including it in double quotation marks:

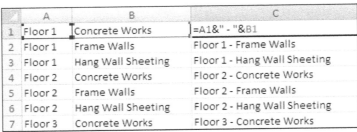

	A	B	C
1	Floor 1	Concrete Works	=A1&" - "&B1
2	Floor 1	Frame Walls	Floor 1 - Frame Walls
3	Floor 1	Hang Wall Sheeting	Floor 1 - Hang Wall Sheeting
4	Floor 2	Concrete Works	Floor 2 - Concrete Works
5	Floor 2	Frame Walls	Floor 2 - Frame Walls
6	Floor 2	Hang Wall Sheeting	Floor 2 - Hang Wall Sheeting
7	Floor 3	Concrete Works	Floor 3 - Concrete Works

 When you copy and paste dates into the schedule, you may find that tasks are assigned constraints, which you may not desire. It is recommended that you display the **Indicators** column, which will show a button if a constraint has been applied. Avoid copying dates.

24.3.3   Dynamically Linking Cells to Other Programs

It is also possible to dynamically link data to other programs such as an Excel spreadsheet:

- Copy the data from the spreadsheet,

- Select the cell position in the table where the data is to be pasted in Microsoft Project,

- Select **Task**, **Clipboard** group, **Paste Special**, **Paste Link**, **Text Data** option,

- The data will be pasted into the cell(s) and changes to linked cells in the spreadsheet or other program will be reflected in the Microsoft Project schedule.

- The linked cell will have a little triangle in the bottom right-hand side indicating that the cell is linked. The Task Name and Duration cells in the picture below are linked.

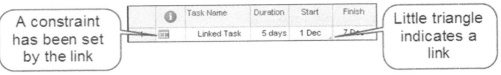

- Be careful when linking dates as this sets constraints. Because the Start date is linked in the picture above this task also has a constraint which is indicated by the ⊞ **Constraint** icon in the indicator column.

- When reopening the project schedule at a later date you will be asked if you wish to refresh the data from the other application.

- To remove a link delete or change the data in the cell.
- To open the linked document, double-click on the little triangle in the bottom right-hand side of the cell.

It is also possible to link one or more cells in a schedule with another cell in the same schedule so a change in one cell will change all the other linked cell(s) using the **Paste Link** option.

 When a field has been linked to another file you must be careful when changing the file names so you do not sever the link.

24.3.4 Milestones with a Duration

Microsoft Project also has the ability to display a Milestone with a non-zero duration. The Milestone duration may be elapsed, 24-hour per day 7-days per week which is created by typing an "e" before the duration units, or a calendar calculated duration. To create a task with a non-zero duration:

- Highlight the task with a duration that you want to mark as a milestone,
- Open the **Task Information** form by double-clicking on the task,
- Select the **Advanced** tab, and
- Click the **Mark task as milestone** check box.

Task Name	Duration	Start	Finish
Milestone	0 days	01 Sep 08:00	01 Sep 08:00
Task 1	4 days	01 Sep 08:00	04 Sep 17:00
Milestone with a duration	4 days	05 Sep 08:00	10 Sep 17:00
Task 2	2 days	11 Sep 08:00	12 Sep 17:00
Milstone with an elapsed duration	4 edays	04 Sep 17:00	08 Sep 17:00
Task 3	2 days	09 Sep 08:00	10 Sep 17:00

The picture above shows that the Milestone point is displayed at the end of the task duration.

 The use of this is not recommended as it makes the schedule difficult to read.

24.3.5 Splitting an In-progress Task

A task may be manually split by clicking on the **Task**, **Schedule** group button, moving your cursor over the point on the task bar where you want a split, and dragging the remaining part of the task. Splitting a task effectively increases the duration of the task but work associated with a split task is scheduled intermittently.

The **File**, **Options**, **Schedule** tab, **Scheduling options for this project:**, **Split in-progress tasks** option does not need to be checked for manual splitting of a task. It must be checked to allow the splitting of a task by Out Of Sequence Progress when a task starts before a predecessor is complete.

An in-progress task may also be split by:

- Dragging the incomplete portion of a task in the bar chart, or
- Using the **Reschedule uncompleted work to start after:** function, or

- Commencing a task before its predecessor finishes as shown by the second task in the diagram below:

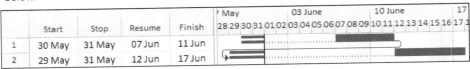

- Task 2 has been split by the relationship with Task 1.

- The Stop and Resume dates, shown in the picture above, may be also manually edited.

24.4   Formatting Bars

24.4.1   Format One or More Specific Task Bars

One or more individual bars may be formatted to make them look different from other bars. To format one or more bars:

- Select:
  - ➢ One task bar by clicking on it, or
  - ➢ Multiple bars by Ctrl-clicking each bar or left-clicking and dragging with the mouse, then

- Open the **Format Bar** form by:
  - ➢ Selecting **Format**, **Bar Styles** group **Format**, **Bar**, or
  - ➢ Moving the mouse over a bar in the bar chart until the mouse changes to a ✥ and double-clicking. When more than one bar has been selected you will need to hold the **Ctrl Key** down when double-clicking on a bar.

- Then select the required bar formatting options from the **Format Bar** form:

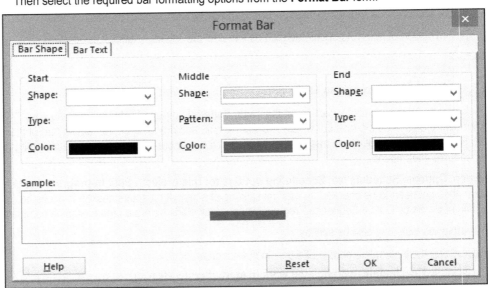

This publication is only sold as a bound book, no parts may be reproduced by any means, e.g. electronic, video or print.

© **Eastwood Harris Pty Ltd**   334

- The [Reset] button is used to restore the default bar formatting to selected bars.

The bar shape may be formatted and text information added in the same way as formatting all the bars described earlier in this chapter.

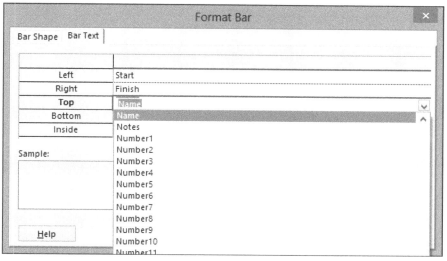

- Select **Format**, **Format** group, **Text styles** to format the text font.

24.4.2 Hiding the Task Bar

A Task Bar may be hidden by checking the **Hide bar** option in the **Task Information** form, **General** tab which is opened by double clicking on the Task Name.

24.4.3 Layout Form – Format Bars Options (Date, Height and Rollup)

Select **Format**, **Format** group, **Layout** to open the **Layout** form. This form has some additional bar formatting options for users to customize the appearance of the Gantt Chart.

- **Date format:** sets the format for dates displayed on bars only. Dates are displayed on bars using **Format**, **Bar Styles** group, **Format** button and selecting either the **Bar** or **Bar Styles** buttons.

- **Bar height:** sets the height of all the bars. Individual bars may be assigned different heights by selecting a bar shape in the styles form.

- **Always roll up Gantt bars** and **Hide rollup bars when summary expanded** works as follows:

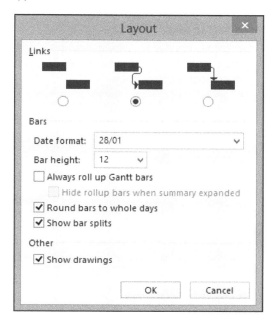

> Tasks before roll up:

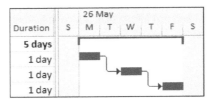

> With **Always roll up Gantt bars** checked and **Hide rollup bars when summary expanded** unchecked:

> With **Always roll up Gantt bars** and **Hide rollup bars when summary expanded** checked:

> An individual bar could be rolled up to a summary task in earlier versions of Microsoft Project using the **Roll up Gantt bar to summary** option in the **Task Information** form when **Always roll up Gantt bars** options are unchecked. This command has been renamed to **Rollup** in Microsoft Project 2013. This function did not work in the author's version of Microsoft Project 2010 but it may have been fixed with an update.

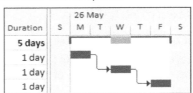

- **Round bars to whole days:**
  > When this option is unchecked, the length of the task will be shown in proportion to the total number of hours worked per day over the 24-hour time span. For example, an 8-hour working duration bar is shown below:

  > When this option is checked, the task bar will be displayed and spanned over the whole day irrespective of working time:

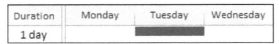

 This function is useful as it make short duration tasks more visible on long schedules.

24.5   Task Splitting

24.5.1   Splitting Tasks

An un-started task may be split using the **Split Task** button 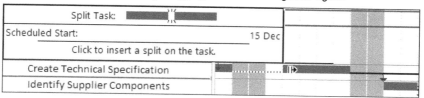 on the **Task**, **Schedule** group, then highlighting the bar to be split and dragging the section of the bar with the mouse to the location where it is planned to conduct the work. The picture below shows the Create Technical Specifications task being split. Splits may be removed by dragging the bar back together again.

In-progress tasks may be split manually by dragging the incomplete portion to the right or a split may be created automatically by commencing a task before its predecessor is complete. Splitting in-progress tasks is covered in both the **TRACKING PROGRESS** and **OPTIONS** chapters.

24.5.2   Show bars splits

The two pictures below are of the same task, first with the option checked and then unchecked:

- When checked, the task bar will display splits:

- When unchecked, the task bar will **NOT** display splits:

24.5.3   Bar Split Dates.

- The start and finish dates of splits of un-started tasks are not available through the user interface, only the start and finish of the task is available. When an in-progress task is split then the **Stop** date and **Resume** date may be seen in columns, see para 24.9.5.

24.5.4   Split Task Duration

The **Duration Type**, which may be displayed in a column or the **Task Information**, **Advanced** tab etc., determines if a split task displays the working duration or the elapsed duration:

- **Fixed Units** and **Fixed Work** show the worked duration of 10 days, and

- **Fixed Duration** shows the duration of 15 days from the start to finish of a task.

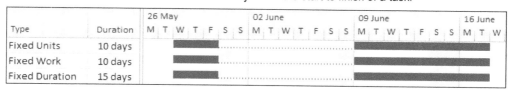

24.6   Logic

24.6.1   Unique ID Predecessor or Unique ID Successor Columns

Each task is assigned a Unique ID when it is created and this number is not used again in the schedule, even if the task is deleted. The Unique ID is changed for all new tasks, even when they are created by copy and pasting existing tasks.

There are two other columns that may be used to edit and display relationships using the Unique ID:

- The **Unique ID Predecessor**, and
- The **Unique ID Successor**.

The Task **Unique ID** will allow users to identify easily which tasks have been added or deleted when a revised schedule has been submitted. On the other hand if one wants to reset the unique ID, or hide the addition or deletion of tasks then a new schedule may be created, the calendars transferred with Organizer and all the tasks copied and pasted into the new schedule. There is also a unique **Resource ID** and a Resource Assignment **Unique ID**.

24.6.2   Reviewing Relationships Using WBS Predecessor or Successor Columns

There are two other columns that may be used only to display (and not edit) the **Predecessors** and **Successors**:

- The **WBS Predecessor**, and
- **WBS Successor**.

24.6.3   Task Drivers

A task may not be on the Critical Path and may have more than one predecessor. A **Driving Relationship** is the predecessor that determines the Early Start of a task.

- Microsoft Project 2000 – 2003 does not identify the difference between **Driving** and **Non-driving Relationships**, which often makes analyzing a schedule difficult. With earlier versions of Microsoft Project often the simplest way to determine the driving relationship for tasks not on the Critical Path and with more than one predecessor, was to delete the relationships until the task moved.

- Oracle Primavera products have always displayed the driving predecessors and successors in the Predecessor and Successors form.

- Microsoft Project 2007 introduced a **Task Drivers** form that indicates which is the driving predecessor and whether the schedule has been Resource Leveled. It also displays the effects of leveling.

In Microsoft Project 2013 the **Task Drivers** form has been renamed **Inspect Task** and additional functionality has been added. Select **Task**, **Tasks** group, **Inspect**, [?] **Inspect Tasks** button open the **Task Inspector** pane:

- The picture below shows that Task 10 is the driving predecessor of Task 11.

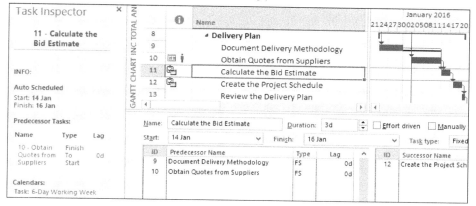

- The picture below shows that Task 15 has issues with resource overloading and makes some suggestions for resolving the issues:

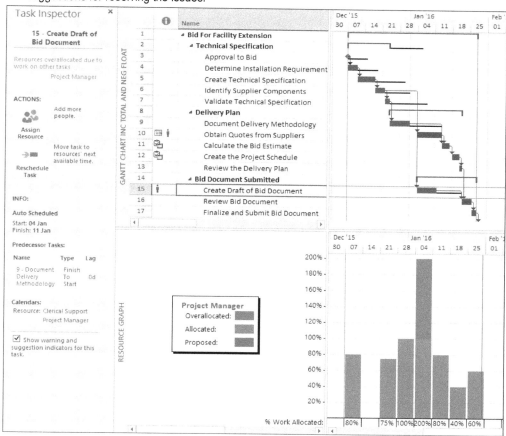

24.6.4 Manually Scheduled Relationships

When two **Manually Scheduled** tasks are linked they acknowledge the relationship when:

- The **File**, **Options**, **Schedule** tab, **Scheduling options for this project:** section, **Update Manually Scheduled tasks when editing links** is checked, and

- A relationship is added or edited.

- In the pictures below:
  - ➢ Tasks 9 has been dragged back in time and the Finish-to-Start relationship is not being acknowledged and
  - ➢ The **Task Inspector**, which may be opened by selecting **Task**, **Tasks** group, **Inspect**, has highlighted the Finish Date of Task 9 and the outline of Task 9 is now dotted. This dotted outline will only display when the **File**, **Options**, **Schedule**, **Schedule Alerts Options**, **Show task schedule warnings** is clicked on:

Before change:

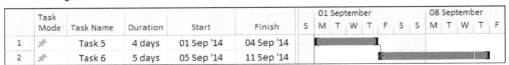

After Change

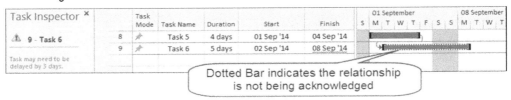

24.6.5 Schedule From Project Finish Date

You are able to impose an absolute project finish date by setting the **Schedule from:** option to **Project Finish Date** in the **Project Information** form.

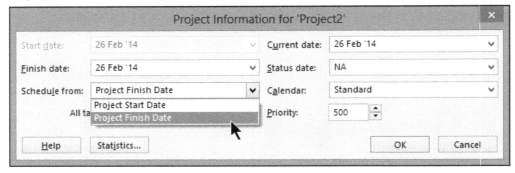

This option may be set after tasks have been added to the schedule. From the point in time that the project is set to schedule from the Project Finish date all new tasks will be assigned with an **As Late As Possible** (ALAP) constraint as they are created.

Any original tasks which may be set as **Soon As Possible** (ASAP) may either be:

- Reset as ALAP, and calculated with all new tasks as ALAP, or

- Left as ASAP.

When the original tasks are left as ASAP, the tasks will be scheduled as ASAP with a Start No Earlier constraint which is calculated either:

- With no float when their combined durations are greater than the ALAP tasks, and the Total Float extends beyond the **Project Finish Date**, identified by the vertical line in the pictures below:

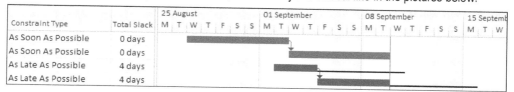

- Will not start earlier than the earliest Early Start of the ALAP tasks when their combined durations are greater than the ALAP tasks.

 Unlike in Oracle Primavera software it is not possible in Microsoft Project to set both a start and finish date on a project. These dates are often called project constraints.

24.6.6 Move Project

The **Project**, **Schedule** group, 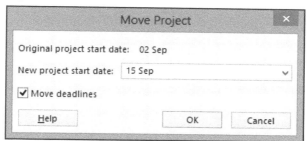 **Move Project** function opens the Move Project form which allows the resetting of the **Project Start** date:

 This option will change Actual Dates and should be used with caution if used with a progressed schedule.

24.7   Custom Outline Codes and WBS

24.7.1   Custom Outline Codes

There are **Task Custom Outline Codes** and **Resource Outline Codes** that may be renamed to suit the project requirements.

- Task Custom Outline Codes may be used for any hierarchical project breakdown structure, such as a PRINCE2 Product Breakdown Structure, Contract Breakdown Structure, and

- Resource Custom Outline Codes may be used for organizational breakdown structures such as the hierarchy of authority, locations and departments.

The process to use this function has three steps:

- Define the new Custom Outline Code structure,

- Assign the codes to the tasks, and

- Create a Group to organize the tasks under the new Custom Outline Code structure.

There is no **Ribbon** menu button for opening the **Custom Fields** form therefore you should consider adding the **Custom Fields** button to your **Quick Access Toolbar** or **Ribbon**.

24.7.2   Define a Custom Outline Code

Click on the **Custom Fields** button

to open the form:

- An Outline Code may be created for either Task or Resource or Project in an Enterprise version data by clicking on the appropriate radio button under the title **Field**.

- Select **Outline Code** from the **Type:** drop-down box,

- The ⎿ Rename... ⏌ button opens a form to edit the name of the Outline Code. Two codes in the picture have been renamed Stage and System.

- The Import Field... opens the **Import Custom Field** form allowing a code structure from another project to be imported in a similar method to Organizer.

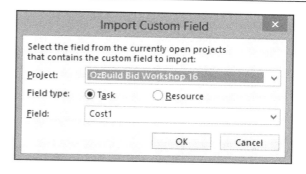

- **NOTE:** Text from another product such as Excel may be simply cut and pasted into most Microsoft Project lists and does not have to be manually typed in.

- The Lookup... opens **Edit Lookup Table** form for the selected Outline Code.

- Before codes are entered the **Mask** or code structure may be defined by clicking on the Edit Mask... button at the top right-hand side. This will open the **Outline Code Definition** form where the code structure is defined.

    ➢ As each **Level** is created it is assigned a number.

    ➢ The **Sequence** defines the type of text that may be entered for the code: Numbers, Uppercase, Lowercase or Characters (text).

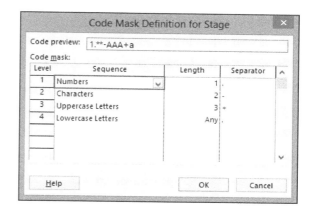

    ➢ The **Length** specifies how many characters the Code Level may have: any number of characters, or a number between 1 and 10.

    ➢ The **Separator** defines the character that separates each level in the structure.

- Click the [OK] button to return to the **Edit Lookup Table** form where the Code Values and Descriptions are entered:

  ➤ The picture shows two "Units" at level 1 and each Unit's Subsystem at level 2.

  ➤ The buttons along the top of the form have a similar function to their use in Outlining and may be used to indent and outdent codes and copy and paste code groups.

  ➤ The other options in the form are self-explanatory.

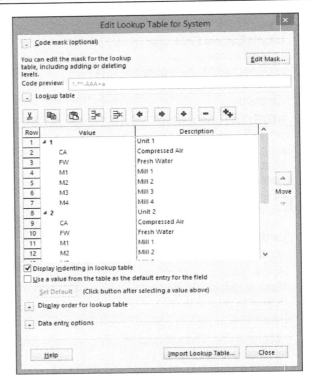

Assigning Custom Codes to Tasks

The codes are assigned by:

- Displaying the appropriate column:

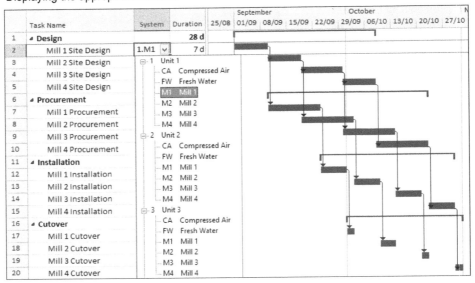

- Or by opening the **Task Information** or **Resource Information** form:

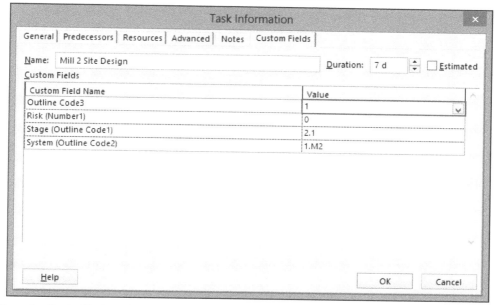

Organize Tasks Under a Custom Outline Code Structure

Select the **View**, **Data** group, **Group by:**, **More Groups...** option to create a grouping to display the tasks under the Custom Outline Code. The same process is used as outlined in the previous section to create a grouping:

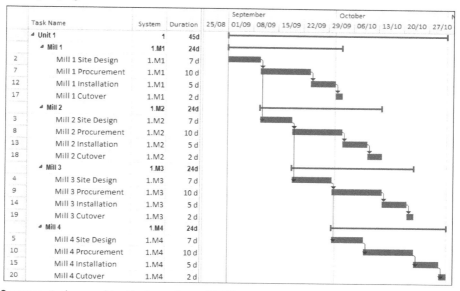

The Summary tasks are virtual tasks and may **NOT** have resources and costs assigned to them.

 There are some restrictions with the use of Custom Outline Codes as the code **Value** is only shown in columns and the code **Description** only in the banding.

24.7.3 Outline Codes

The Microsoft Project **Outlining** function may be used to code the Project Breakdown Structure.

- Display the Outline Code against each task in the schedule by displaying the **Outline Number** column.

- An **Outline Level** displays how many levels down in the Outline Code structure the Task lays.

- The **Outline Number** may be displayed against the task Name by adding the [Outline Number] button to the Quick Access Toolbar.

	Outline Number	Outline Level	Name
1	1	1	⊿ **1 Bid For Facility Extension**
2	1.1	2	⊿ **1.1 Technical Specification**
3	1.1.1	3	1.1.1 Approval to Bid
4	1.1.2	3	1.1.2 Determine Installation Requirements
5	1.1.3	3	1.1.3 Create Technical Specification
6	1.1.4	3	1.1.4 Identify Supplier Components
7	1.1.5	3	1.1.5 Validate Technical Specification
8	1.2	2	⊿ **1.2 Delivery Plan**
9	1.2.1	3	1.2.1 Document Delivery Methodology
10	1.2.2	3	1.2.2 Obtain Quotes from Suppliers
11	1.2.3	3	1.2.3 Calculate the Bid Estimate
12	1.2.4	3	1.2.4 Create the Project Schedule
13	1.2.5	3	1.2.5 Review the Delivery Plan
14	1.3	2	⊿ **1.3 Bid Document Submitted**
15	1.3.1	3	1.3.1 Create Draft of Bid Document
16	1.3.2	3	1.3.2 Review Bid Document
17	1.3.3	3	1.3.3 Finalize and Submit Bid Document
18	1.3.4	3	1.3.4 Bid Document Submitted

24.7.4 User Defined WBS Function

There are occasions when information must be presented under a predefined WBS Code structure. This structure could be a company standard, project-specific need, or defined by a client. The **WBS Code Definition** function is a method of tailoring the display of the **Outline Code** in a column titled **WBS**. This does not provide an additional set of codes to organize your project tasks; it only allows an alternate display of the Outline Code.

In a new project, the default WBS code is identical to the Outline Code. The example below displays both the WBS and Outline codes of the OzBuild schedule before doing any tailoring of the WBS code:

	WBS	Outline Number	Outline Level	Name
1	1	1	1	⊿ **1 Bid For Facility Extension**
2	1.1	1.1	2	⊿ **1.1 Technical Specification**
3	1.1.1	1.1.1	3	1.1.1 Approval to Bid
4	1.1.2	1.1.2	3	1.1.2 Determine Installation Requirements
5	1.1.3	1.1.3	3	1.1.3 Create Technical Specification
6	1.1.4	1.1.4	3	1.1.4 Identify Supplier Components
7	1.1.5	1.1.5	3	1.1.5 Validate Technical Specification
8	1.2	1.2	2	⊿ **1.2 Delivery Plan**
9	1.2.1	1.2.1	3	1.2.1 Document Delivery Methodology
10	1.2.2	1.2.2	3	1.2.2 Obtain Quotes from Suppliers
11	1.2.3	1.2.3	3	1.2.3 Calculate the Bid Estimate

The WBS Codes structure may be tailored to suit your own requirements. Below is an example of a WBS code that has been tailored using the WBS code function. The Outline sequential numbers may be replaced by user defined sequences of numbers or letters.

	WBS	Outline Number	Name
1	**OzBuild1**	**1**	◢ **Bid For Facility Extension**
2	**OzBuild1.AA**	**1.1**	◢ **Technical Specification**
3	OzBuild1.AA.001	1.1.1	Approval to Bid
4	OzBuild1.AA.002	1.1.2	Determine Installation Requirements
5	OzBuild1.AA.003	1.1.3	Create Technical Specification
6	OzBuild1.AA.004	1.1.4	Identify Supplier Components
7	OzBuild1.AA.005	1.1.5	Validate Technical Specification
8	**OzBuild1.AB**	**1.2**	◢ **Delivery Plan**
9	OzBuild1.AB.001	1.2.1	Document Delivery Methodology
10	OzBuild1.AB.002	1.2.2	Obtain Quotes from Suppliers
11	OzBuild1.AB.003	1.2.3	Calculate the Bid Estimate
12	OzBuild1.AB.004	1.2.4	Create the Project Schedule
13	OzBuild1.AB.005	1.2.5	Review the Delivery Plan
14	**OzBuild1.AC**	**1.3**	◢ **Bid Document Submitted**
15	OzBuild1.AC.001	1.3.1	Create Draft of Bid Document
16	OzBuild1.AC.002	1.3.2	Review Bid Document
17	OzBuild1.AC.003	1.3.3	Finalize and Submit Bid Document
18	OzBuild1.AC.004	1.3.4	Bid Document Submitted

To create your own user defined WBS codes, you will need to:

- Add the **WBS** button on your **Ribbon** or **Quick Access Toolbar**,

- Click on the ⊟ **WBS** button,

- Click on the ⊞ **Define Code** button to open the **WBS Code Definition** form:

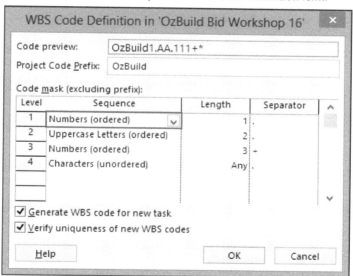

- The **Project Code Prefix:** is where you enter a text string that will precede all **WBS** codes. See the example above.

- Each code level is defined in the **Sequence** column. There are four code types. Each is displayed in the picture above:

  ➢ Numbers (ordered)

  ➢ Uppercase Letters (ordered)

  ➢ Lowercase Letters (ordered)

  ➢ Characters (unordered)

  The **Numbers**, **Uppercase Letters** and **Lowercase Letters** are numbered automatically.

  The **Characters** option will initially define all WBS codes at this level as an * and these may be overwritten with text. Microsoft Project will not renumber them.

- A **Length** of 1 to 10 or "Any" may be selected as the length of the code.

- Check **Generate WBS code for new task** and each new task will be automatically coded when created. When left unchecked, no WBS code is assigned to new tasks.

- Check **Verify uniqueness of new WBS codes** to prevent the creation of duplicate WBS codes.

You may renumber the WBS codes y selecting **All** or **Selected Tasks**, Select **Project**, **Properties** group, **WBS**, **Renumber...** from the menu.

24.8 *Sharing Resources with Other Projects*

Microsoft Project allows resources to be shared with multiple projects. This feature is not covered in detail in this book. Select **Resource**, **Assignments** group, **Resource Pool**, **Share Resources...** to open the **Share Resources** form, select the project to share resources from and set calculation options for leveling.

24.9 Tracking Progress

24.9.1 Setting an Interim Baseline

It may be necessary to save an interim baseline. This may occur when the scope of a project has changed and a new baseline is required to measure progress against, but at the same time you may also want to keep a copy of the original baseline. This process may also be used to display the effect of scope changes on an original project plan by comparing one baseline with another.

There are two types of data fields to save interim baselines:

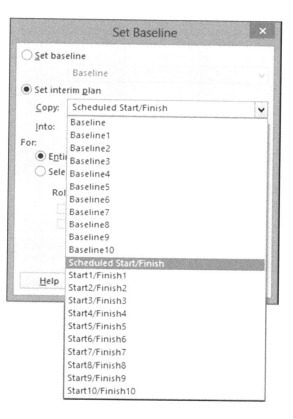

- Using one of the 10 additional baselines titled **Baselines 1** to **10**. These Baselines will save start date, finish date, work and cost information.

- Using one of the 10 sets of **Start Date**, **Finish Date** and **Duration** fields. This function will only save the start and finish date information. This is termed an **Interim Plan** by Microsoft Project.

- **Set baseline** to save the current schedule to one of the 11 Baselines, or

- **Set interim plan** then:
    - ➤ Select the dates you want to copy from using the **Copy:** drop-down box, and then
    - ➤ Select where you want to copy the dates to using the **Into:** drop-down box.

 The **Baseline** data may be reviewed in some Views such as the **Task Details Form** and in columns. You will be able to display the **Baseline 1** to **10** and **Interim Plan** dates and durations in columns and as a bar on the Gantt Chart but not in the forms. Therefore, it is recommended that the current baseline be saved as **Baseline** since the data is more accessible that way. Previous baselines should be copied to **Baselines 1** to **10**.

Custom Date fields may be renamed with the **Customize Fields** form. Select **Project**, **Properties** group, **Customize Fields** to open the **Customize Fields** form. This may be useful to record the reason why a baseline has been saved.

24.9.2 Resetting the Baseline Using "Roll Up Baselines"

When a detail task is moved from one Summary task to another, the Baseline dates are moved with the detail task. The Summary task Baseline dates are not recalculated so they may not be valid. Likewise, the original Summary Baseline dates are not recalculated when new tasks are added to a schedule so they may not be valid.

When a task has costs or work assigned to it, then any movement of a detail task from one Summary task to another Summary task should result in a change to the baseline value of the Summary tasks, but these new values are not automatically recalculated by Microsoft Project. Costs and Work Baseline calculations are covered in the **UPDATING PROJECT WITH RESOURCES** chapter.

In the pictures below, the task Detailed 2.3 was moved from below Parent 2 to below Parent 1 so the baseline of tasks Parent 1 and Parent 2 (the lower bar on each task) no longer accurately reflect the summary of the detail task Baseline dates.

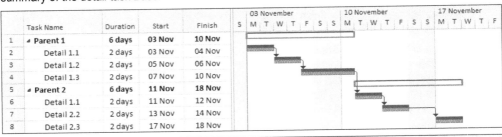

 In the example above, when the task was moved the option **Autolink inserted or moved tasks** found under the **File**, **Options**, **Schedule** tab is unchecked, otherwise the logic would have been changed. The formatting of these bars has been achieved using the Gantt Chart Wizard which results in clearer screen shots.

The **Roll up baselines** option in the **Set Baseline** form will recalculate the summary task Baseline data when a detail task is moved to a different Summary task or a new task is added to the schedule. There are three options when **Selected tasks: - Roll up Baselines:** is selected:

- Check only **To all summary tasks**,

- Check only **From subtasks to selected summary task(s)**, or

- Check both of the above options.

 The options function differently when a Summary task or when a detail task is selected. These options are difficult to understand and the author prefers to reset Baselines of Summary tasks manually.

In the example below two new tasks have been added, thus the old Baseline is no longer the same as the current schedule. In addition the start date has been delayed:

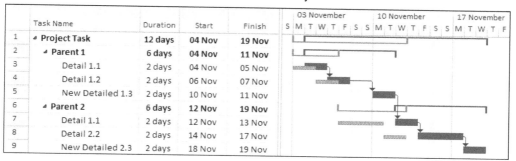

To all summary tasks

- When the **To all summary tasks** option is checked and only a **Detailed task** is selected, such as the New Detailed 2.3 task below, then in this option:

  ➢ The selected detail task has its Baseline set based on the current schedule, see New Detailed 2.3 task, and

  ➢ Related Summary tasks have their baselines set based on the new detail task Baseline, see Parent Task and Parent 2.

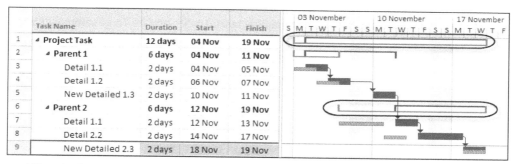

- When the **To all summary tasks** option is checked and a **Detailed task** and a **Summary task** is selected, such as the Parent 2 and New Detailed 2.3 task below, then:

  ➢ The selected detail task and summary task have their Baseline set based on the Current dates, not the Baseline dates, see New Detailed 2.3 task and Parent 2,

  ➢ Related summary tasks have their baselines set based on the new Baseline dates, see Project Task:

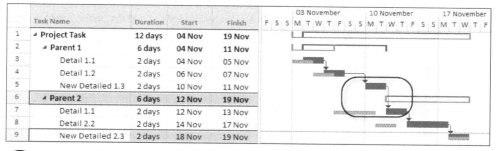

 Resetting Baseline dates of the summary task may not be intended and in the example above the Parent 2 Baseline dates are no longer summarized on the detail tasks Baseline dates. So when using the **To all summary tasks** option it is recommended that no summary tasks should be selected.

From subtasks to selected summary task(s)

- When the **From subtasks to selected summary task(s)** option is checked and both Summary and detail tasks are selected, such as Parent 2 and New Detailed 2.3 tasks below:
  - ➤ The selected detail tasks have their Baseline reset based on the current schedule,
  - ➤ Selected summary tasks have their Baseline dates set based on the Baseline dates of their detail tasks,
  - ➤ The unselected summary tasks do not have their baselines reset, see task Parent 1 and Parent Task below.

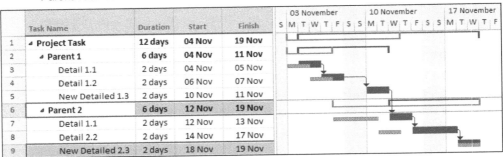

From subtasks to selected summary task(s) and To all summary tasks option selected

- When the **From subtasks to selected summary task(s)** option and the **To all summary tasks** option is checked then:
  - ➤ Any selected detail task has the Baseline set or reset, and
  - ➤ All associated summary tasks have their Baselines reset based on the associated Baseline dates.

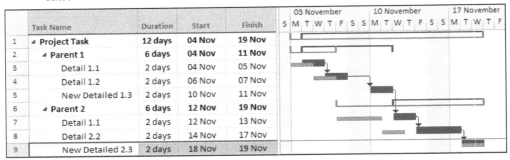

 It is recommended to set a revised baseline when a new detail task has been added. The new task and the associated Parent tasks should be highlighted and the option **From subtasks to selected summary task(s)** applied.

24.9.3 Status Date Calculation Options - New Tasks

New functions were introduced in Microsoft Project 2002 intended to assist schedulers to place the new tasks as they are added to the schedule in a logical position with respect to the **Status Date**. If the **Status Date** has not been set then the **Current Date** is used. In the example below the vertical black line is the **Status Date**:

- Task 1 has completed work in the future, which is not logical,
- Task 2 would normally be considered in the logical position with respect to the **Status Date**, and
- Task 3 has incomplete work in the past, which is also not logical.

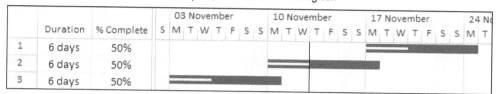

These options are found under **File**, **Options**, **Advanced** tab, **Calculation options for this project:** section:

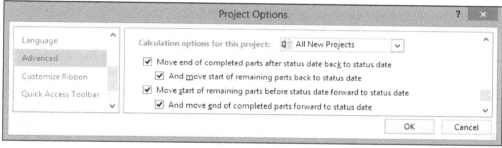

For all these options to operate all four of the following parameters must be met:

- The **Split in-progress tasks** option in the **Schedule** tab must be checked, and
- The required option on the **Calculation** tab must be checked before the task is added or edited, and
- The **Updating task status updates resource status** option on the **Calculation** tab must be checked.
- The Task **MUST NOT BE** assigned **Task Duration Type** of **Fixed Duration**, which is often recommended.

 The documentation found in the earlier Microsoft Project Help files is not precise and does not make it clear to the user that these options may NOT be turned on and off to recalculate all tasks. The options only work on new tasks when they are added to a schedule or when a task is updated by changing the % Complete.

The following four options are available:

- Check only **Move end of completed parts after status date back to status date**. This is the effect on Task 1 in the picture on the previous page when the % Complete is increased to 60%:

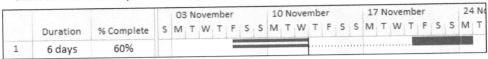

	Duration	% Complete
1	6 days	60%

- The **And move start of remaining parts back to status date** option is only available after the option above is checked, if checked the result would be as per the picture below:

	Duration	% Complete
1	6 days	70%

- The **Move start of remaining parts before status date forward to status date** may be checked at any time. The picture below shows the effect on Task 3 from the picture on the previous page when the % Complete is changed to 60%:

	Duration	% Complete
3	6 days	60%

- The **And move end of completed parts forward to status date** may be checked after the option above is checked and before a % Complete is entered, if checked the result would be as per the picture below on Task 2:

	Duration	% Complete
2	6 days	40%

This function will ignore constraints even when the Schedule Option **Tasks will always honor their constraint dates** has been set.

 Experimentation by the author indicates that the tasks will adopt the rules that are set in the options form when the task is added or when a task % Complete is changed. Therefore this function may not be applied to existing schedules, but only to new tasks if the options are set before the tasks are added or when a task % Complete is updated.

This function in its current form has some restrictions that schedulers may find unacceptable:

- Existing schedules may not be opened and the function applied.
- When the **Move start of remaining parts before status date forward to status date** is used, it overrides any **Actual Start** date that you have entered prior to entering a % Complete.

This option should be used with caution and users should ensure they fully understand how this function operates by updating a simple schedule multiple times.

24.9.4   Status Date Calculation Options - When Updating a Schedule

The Status Date Calculation Options also operates when a schedule is updated by changing the % Complete of Tasks. The example below shows a task with the Status Date, the dark vertical line, set to the next period:

When all the **Status Date Calculation Option** options are checked and the percent complete is changed from 0% to 40% the task is aligned to the new Status Date:

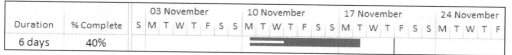

Now the Status Date has been moved to the next week. The tasks have not moved in respect to the Status Date and will not move when the project is recalculated:

Duration	% Complete																									
6 days	40%																									

After percent complete is changed to 60%, the task splits and the un-worked portion of the task is set to be completed after the new Status Date:

Duration	% Complete																									
6 days	60%																									

24.9.5   Stop and Resume Dates

The **Stop** date represents the point in time where the completed work finishes. The **Resume** date represents the point in time that the incomplete work is scheduled to restart. These dates may be displayed in columns and edited from these columns. Task B Resume date has been manually edited:

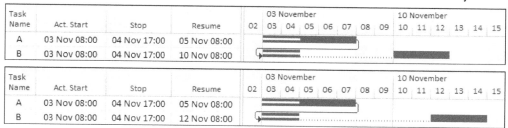

24.9.6 Marking Up Summary Tasks

It is not normal to mark up summary tasks and is not recommended. Actual Dates may not be entered against summary tasks. But a % Complete may be entered against a summary task and all the subordinate tasks will be auto-updated to match the summary % Complete.

When a **% Complete** is entered against a summary task, all subordinate tasks, both Summary and Detail, inherit a value depending upon the setting of the **Updating task status updates resource status** option found in **File**, **Options**, **Schedule** tab, **Calculation options for this project:** section.

To explain this concept further, see the following two examples. Both schedules pictured below had 15% entered against Task 1, the OzBuild Initiation task. All other Actual dates and % Completes were calculated by Microsoft Project.

- With the **Updating task status updates resource status CHECKED** the tasks are updated as if they were completed according to plan:

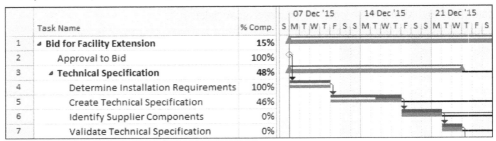

- With the **Updating task status updates resource status UNCHECKED** all tasks are assigned to 15% complete, which is not the logical progression of the project:

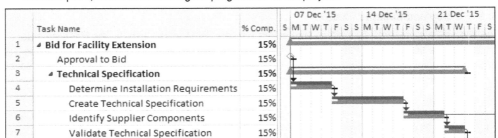

24.9.7 Moving Tasks Using Task, Tasks group Move

The **Task**, **Task** group **Move** functions were new to Microsoft Project 2010. They are able to move a task backward or forward in time:

To demonstrate some of the functions of these commands a task has been added to a schedule and positioned on the Project Start Date:

- Moved unprogressed tasks will have a constraint set without warning and this is similar to dragging a task which will also set a constraint:

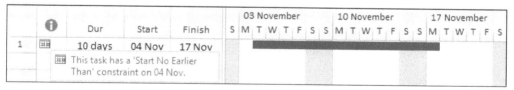

- If it is required to move the incomplete portion of a task then **File**, **Options**, **Schedule** tab, **Scheduling options for this project:** section, **Split in-progress tasks** must be checked otherwise the task will not split. The Task below was moved forward one day after a % Complete had been assigned and once split, it developed an interesting error message:

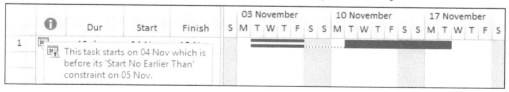

24.9.8 Progress Lines and Variance Columns

Start Variance and **Finish Variance** columns are variances from the Baseline and may be displayed in columns as per the picture below.

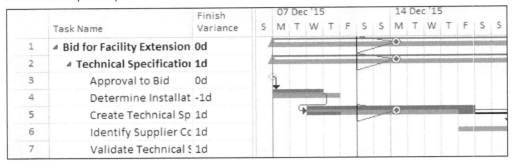

Some users like to display **Progress Lines** which are usually displayed as zigzag lines on the Gantt Chart showing how far ahead or behind the project tasks are:

- The **Progress Lines** button should be added to the **Quick Access Toolbar** or **Ribbon**

- Click on the **Progress Lines** to open the **Progress Line** form where the progress lines may be formatted:

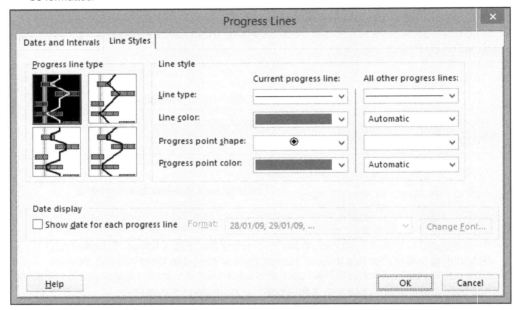

> ℹ Progress Lines become less relevant when a schedule is updated using Actual dates and remaining durations.

25 TOOLS AND TECHNIQUES FOR SCHEDULING

25.1 Understanding Menu Options

You will find that the menu options change as you select rows, cells, all rows and cells, or views. Therefore, not all menu options are always available and some buttons are grayed out and inoperable. The following topics will be covered in this chapter:

- Menu items sometimes found by right-clicking:
  - ➢ **Cut**, **Copy** and **Paste Task**
  - ➢ **Cut**, **Copy** and **Paste Cell**
  - ➢ **Copy Picture**
  - ➢ **Fill**
  - ➢ **Clear**
  - ➢ **Find** and **Replace**
  - ➢ **Go To**
- **Insert**, **Recurring tasks...**
- **Splitting** a Task
- **Copy**, **Cut** and **Paste** from Spreadsheets
- **Unique Task**, **Resource** and **Assignment ID**
- **Organizer** is used to transfer data such as Tables and View from one project to another.
- **File types**

25.2 Cut, Copy and Paste Row

This function allows you to select one or more consecutive or non-consecutive tasks (using Ctrl-click) and either copy them, or cut them and paste as a group to a new location that you select with your mouse. Any inter-task relationships are also copied but external relationships to unselected tasks are not copied. Ensure you select the whole task by highlighting the Task ID.

25.3 Cut, Copy and Paste Cell

You may cut or copy information and then paste into one or more cells from:

- One cell, or
- Adjacent cells by dragging in rows and/or columns, or
- A non-contiguous group of cells by Ctrl-clicking.

This function operates in a similar way to Excel's copy/cut-and-paste operation:

- Highlight the cell(s) you want to copy or cut.
- Select **Copy Cell** or **Ctrl+C** to copy a cell or **Cut Cell** or **Ctrl+X** to cut a cell.
- Position the mouse where you want to paste the data and select **Paste** or **Ctrl+V**.

This is an interesting function since it allows you to copy from one column to another when the format is compatible. This function may also be used for transferring and/or updating your schedule from other software such as Excel which is covered later in this chapter.

25.4   Copy Picture

This allows a section or total screen to be copied to the clipboard and then either pasted into a document or saved as a **gif** file. Select **Task**, **Clipboard** group, select the **Copy** ▤ Copy ˅ button and select ▤ **Copy Picture** to open the **Copy Picture** form:

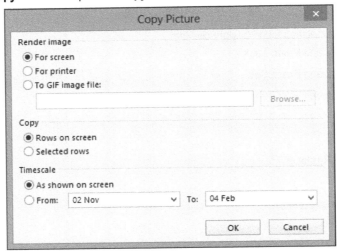

The options are self-explanatory and when pasted into a document the picture looks like the example below:

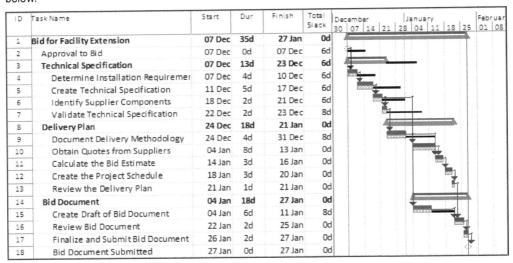

ID	Task Name	Start	Dur	Finish	Total Slack
1	**Bid for Facility Extension**	07 Dec	**35d**	27 Jan	**0d**
2	Approval to Bid	07 Dec	0d	07 Dec	6d
3	**Technical Specification**	07 Dec	**13d**	23 Dec	**6d**
4	Determine Installation Requiremer	07 Dec	4d	10 Dec	6d
5	Create Technical Specification	11 Dec	5d	17 Dec	6d
6	Identify Supplier Components	18 Dec	2d	21 Dec	6d
7	Validate Technical Specification	22 Dec	2d	23 Dec	8d
8	**Delivery Plan**	24 Dec	**18d**	21 Jan	**0d**
9	Document Delivery Methodology	24 Dec	4d	31 Dec	8d
10	Obtain Quotes from Suppliers	04 Jan	8d	13 Jan	0d
11	Calculate the Bid Estimate	14 Jan	3d	16 Jan	0d
12	Create the Project Schedule	18 Jan	3d	20 Jan	0d
13	Review the Delivery Plan	21 Jan	1d	21 Jan	0d
14	**Bid Document**	04 Jan	**18d**	27 Jan	**0d**
15	Create Draft of Bid Document	04 Jan	6d	11 Jan	8d
16	Review Bid Document	22 Jan	2d	25 Jan	0d
17	Finalize and Submit Bid Document	26 Jan	2d	27 Jan	0d
18	Bid Document Submitted	27 Jan	0d	27 Jan	0d

 The author found that not all of the columns were copied, however if the order of the columns was changed then they were all copied. This issue may be fixed with a future update of the software.

25.5   Fill Down

The **Fill Down** command allows you to select one cell or a range of cells and then copy them up, down, left, or right without the need to copy and paste.

25.6   Clear Contents

The **Clear Contents** command allows you to clear the contents from a task.

25.7   Find and Replace

The **Ctrl+F**, **Find** and **Ctrl+H**, **Find** and **Replace** functions allow you to find any task by matching data with your defined criteria. This will also allow you to find the data and then replace it with another piece of information.

25.8   Go To

The **Go To** function allows you to quickly find a date and move the timescale horizontally to that date or to a task if you know the Task ID number. The command would need to be added to the **Quick Access Toolbar** or **Ribbon** to be used:

25.9   Insert Recurring Task

The **Task**, **Insert** group, **Task**, Recurring tasks... function allows the insertion of more than one task that occurs on a regular basis. The options are self-explanatory and are very handy for scheduling meetings and other regular tasks. These tasks use the Rollup bars in the **Bars Styles** form.

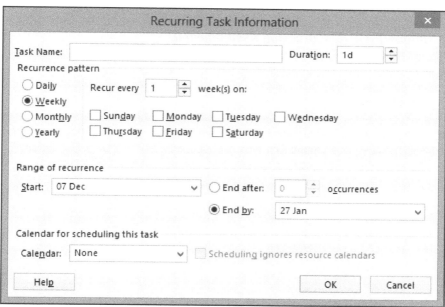

25.10 Copy or Cut-and-Paste to and from Spreadsheets

Microsoft Project will allow **Copy** or **Cut**-and-**Paste** to and from spreadsheets and other software packages. Microsoft Project 2013 will now copy the formatting and headers. This may be useful for a number of purposes:

- Importing Tasks from other applications,
- Assigning Codes and Resources to Tasks,
- Updating a schedule, and
- Exporting Data to other packages.

The **Copy** or **Cut** function copies the task data from highlighted columns and rows in the schedule to the Windows clipboard. These may then be pasted into another software package:

- Highlight the data in your schedule that you want to transfer to another software package and cut or copy the data.
- Move to the spreadsheet application and paste the task information. In earlier versions of Microsoft Project the column headers were not pasted into the application; only data from the schedule was pasted. Microsoft Project 2013 now keeps the formatting and headers.
- The data may be edited or updated in the application, as required.
- The tasks may be selected and pasted back into the schedule to update it with the normal copy and paste functions.

25.11 Paste Link – Cell Values in Columns

A cell may be copied, and then "Paste Linked" to another cell or an external application. Thus when the copied cell is updated, the linked cell is also updated with the same value. For example this could be used to update a number of tasks that have the same % Complete. A linked cell has a small triangle in the corner, and the linking may be removed by overtyping the cell value.

25.12 Unique Task, Resource and Assignment ID

The Task IDs change for all tasks below an inserted task. There are many contractual and management reasons to be able to identify each task by a unique task number.

25.12.1 Task Unique ID

When a new task is added to a blank schedule, it is assigned a **Unique ID** commencing with the number 1. When tasks are deleted these numbers are not re-used and any new task is assigned a new sequential number. This **Unique ID** number may be displayed in a column titled **Unique ID**. It is also possible to display the **Unique Predecessor** and **Unique Successor** columns:

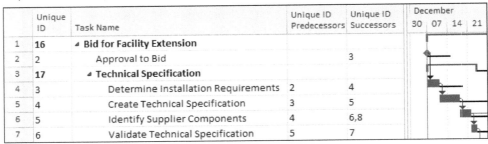

The cutting and pasting of tasks creates new Unique IDs for each task but dragging task to a new position does not. To renumber the Unique IDs just copy all the tasks and paste below the existing tasks and delete the original tasks.

25.12.2 Resource Unique ID

A resource **Unique ID** is created for each resource. Again, when a resource is deleted, the resource **Unique ID** is not re-used. This number may be displayed in the **Resource Sheet** or **Resources Usage** view column titled **Unique ID**.

25.12.3 Resource Assignment Unique ID

When a resource is assigned to a task, the assignment is given a unique **Assignment ID**. This number may be displayed in the **Resource Sheet** or **Resource Usage** view column titled **Unique ID**:

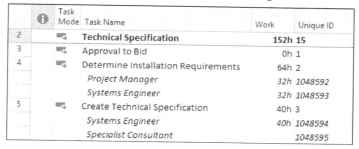

25.13 Organizer

The **Global.mpt** file holds the schedule's default settings such as **Tables** and **Views**, which are inherited by new projects not created from a template. The **Organizer** function is also used to copy information between projects or to update the **Global.mpt**.

The **Organizer** form may be opened from many menus:

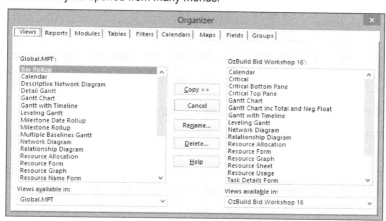

Except for the **Global.mpt**, the projects you want to copy settings to and from will have to be opened in order to copy data.

- The Organizer function is used for copying, renaming and deleting most items such as **Tables**, **Views and Calendars**.

- The two tab titles above that are not self-explanatory are:

  ➢ **Maps** – These are predefined tables for mapping data to be exported or imported, and

  ➢ **Modules** – These are Visual Basic Macros.

25.14 File Types

Microsoft Project is compatible and will operate with the following file types:

- **Microsoft Project (*.mpp)**. This is the default file format for Microsoft Project 2010, 2013 & 2016. All earlier versions will not open or save to a **Project (*.mpp)** file created by Microsoft Project 2010,2013 and 2016.

- **Microsoft Project 2007 (*.mpp)**. This is the default file format for Microsoft Project 2007. This is a different format than the ***.mpp** file created by Microsoft Project 2000 - 2003 and 98.

- **Microsoft Project 2000 – 2003 (*.mpp)**. This is the default file format for Microsoft Project 2000, 2002 and 2003. This is a different format than the ***.mpp** file created by Microsoft Project 2013, 2007 and Microsoft Project 98.

- **Microsoft Project 98 (*.mpp)**. This is the format created by Microsoft Project 98.

  ➢ Microsoft Project 98 will not open or save a **Project (*.mpp)** file created by Microsoft Project 2000 and later versions.

  ➢ Microsoft Project 2000, 2002 and 2003 will open and save to a **Microsoft Project 98 (*.mpp)** file.

  ➢ Microsoft Project 2007, 2010, 2013 & 2016 will not save to a **Microsoft Project 98 (*.mpp)** file.

- **MPX (*.mpx)**. This is a text format data file created by Microsoft Project 98.

  ➢ This format may be opened by all Microsoft Project 2000 - 2003, 2007, 2010 Versions but not with 2013 or 2016. A *.mpx file cannot be created by Microsoft Project 2000 - 2003, 2007, 2010, 2013 & 2016.

  ➢ **mpx** is a format that may be imported and exported by many other older project scheduling software packages, but the trend is to move towards using Microsoft XML format for importing Microsoft Project files into other software packages such as Asta Powerproject and Oracle Primavera P6.

  ➢ Some third-party software will convert mpx files to and from Microsoft Project 2000 – 2003 mpp format files. You may search the Internet for the latest available products.

- **Project Template (*.mpt)**. This 2010, 2013 and 2016 format is used for creating project templates.

- **Microsoft Project 2007 Templates (*.mpt)**. This is the 2007 template format.

- **Project Database (*.mpd)**. This is a Microsoft Project database format, but is not available in Microsoft Project 2007 or 2010.

- **Microsoft Access Database (*.mdb)**. This is the Microsoft Access format in Microsoft Project 2000 – 2003 that is not available in Microsoft Project 2007 and later.

- Data may be saved to (and imported from) files in the following additional formats using **File**, **Save**, **File**, **Save As** and **File**, **Open**:

  ➢ Portable Document Format read by software such as Adobe Acrobat, (*.pdf)

  ➢ XML Paper Specification, a file format developed by Microsoft providing device-independent documents appearance and provides a similar outcome to a pdf file, (*.xps)

  ➢ Excel Workbook (*.xlsx)**)**.

  ➢ Excel Binary Workbook (*.xlsb)

  ➢ Excel 97 - 2003 Workbook (*.xls)

  ➢ Text – Tab delimited text files (*.txt)

  ➢ CSV – Comma delimited text files (*.csv)

➢ XML format (*.xml). Introduced in Microsoft Project 2002, this enables files to be saved in XML (eXtended Markup Language) format allowing data to be shared with other applications. This is becoming the more popular format for importing Microsoft Project files into other software packages such as Asta Powerproject and Oracle Primavera P6.

- The Excel Pivot Table (save only) format from Microsoft Project 2007 is no longer available.

25.15 Recording a Macro

Macros record keystrokes and are useful for performing functions where there are not inbuilt Microsoft Project functions. A typical example is recording a macro to add a Negative Float or Total Float bar. To record a macro:

- Practice the keystrokes you wish to record

- Select Developer,  Record Macro to open the **Record Macro** form:

- Enter the Macro name: without spaces,

- Enter the Shortcut key:, you must not use a letter that has already been assigned such as the letters C, V, Y, B etc.

- Click ⬛ OK ⬛,

- Record the macro by typing your keystrokes,

- When complete select ▪ Stop Recording from the Developer, Code group.

- Press Ctl+ your short cut key to run the macro.

- Select Developer, Code, View Macros ⬛ Options... ⬛ to edit the shortcut key.

26   APPENDIX 1 – SCREENS USED TO CREATE VIEWS

Microsoft Project has the following methods of accessing the data on screen.

- The screen may be split with a **Top** and **Bottom Pane**.
  - ➤ The **Top Pane** displays all project information, except when a filter is applied.
  - ➤ The **Bottom Pane** displays information about the task or tasks or resources that are highlighted in the Top pane.

- A **View** may be created and edited from the **More Views** form. A **View** is created from a **Screen** in the **View Definition** form by clicking [New...] in the **More Views** form.
  - ➤ A **Single View** may be applied to either the **Primary View** (Top Pane) or **Details Pane** (Bottom Pane) as long as the **Screen** may be displayed in that **Pane**.
  - ➤ A **Combination View** selects an existing view for both the **Primary View** (Top Pane) or **Details Pane** (Bottom Pane).

- There are many **Screens**, most of which may be applied to the **Primary View** (Top Pane) or **Details Pane** (Bottom Pane) through **Views**.
  - ➤ There are some restrictions. For example, the **Calendar** screen may not be applied to the bottom pane.
  - ➤ Some **Screens** are not useful when displayed in the incorrect pane. It is better to display the **Task** form in the **Details Pane** and leave a **Gantt Chart** in the **Primary View** (Top Pane).
  - ➤ Some **Screens** are very similar to each other, such as the **Task** screen and the **Task Details** screen.

- Some **Screens** have **Details** forms where the content of a **View** may be changed. An example of this is a **Task Usage** view, which may be formatted to show **Work** or **Costs** and **Cumulative** or **Period**.

- Some Screens are split vertically and have a left-hand side and a right-hand side such as the **Gantt Screen**. Other **Screens** do not split vertically, like the **Task Screen** and the **Task Details Screen**, which are used in the **Task Form** view and the **Task Details Form** view.

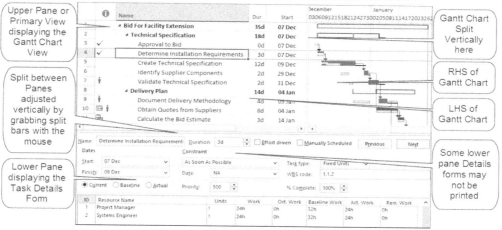

The following table summarizes the **Screen**, **View** and **Details** options:

	Screen Name	Panes	Split Verti-cally	Details Options	Format Column Options
1	Calendar	Top Only	No	None	Right-click Format Options → Go To... Timescale... Gridlines... Text Styles... Bar Styles... Zoom... Layout... Layout Now Show Timeline ✓ Show Split
2	Gantt Chart	Best in Top	Yes	Bars and Columns may be formatted	Yes
3	Network Diagram	Best in Top	No	Task Boxes may be edited	No
4	Relationship Diagram	Either	No	None	No
5	Resource Form	Best in Bottom	No	✓ Show Split Schedule Work Cost Notes Objects	No
6	Resource Graph	Best in Bottom	No	Gridlines... Bar Styles... Peak Units Work Cumulative Work Overallocation Percent Allocation Remaining Availability Cost Cumulative Cost Work Availability Unit Availability Show Timeline ✓ Show Split	No

	Screen Name	Panes	Split Verti-cally	Details Options	Format Column Options
7	Resource Name Form	Best in Bottom	No	☑ Show Split / 🗓 Schedule / 🖨 Work / Cost / Notes / Objects	No
8	Resource Sheet	Either	No	None	Yes
9	Resource Usage	Either	Yes	Detail Styles... / ✓ Work / Actual Work / Cumulative Work / Overallocation / Cost / Remaining Availability / Show Timeline / ✓ Show Split	No
10	Task Details Form	Best in Bottom	Yes	✓ Show Split / Predecessors & Successors / Resources & Predecessors / Resources & Successors / Schedule / Work / Cost / Notes / Objects	No
11	Task Form	Best in Bottom	No		No
12	Task Name Form	Best in Bottom	No		No
13	Task Sheet	Best In Top	No	None	Yes
14	Task Usage	Best in Bottom	Yes	Detail Styles... / ✓ Work / Actual Work / Cumulative Work / Baseline Work / Cost / Actual Cost / Show Timeline / ✓ Show Split	Yes
15	Team Planner – Professional Version Only	Top Only	No	✕ Delete / Information... / Auto Schedule / Manually Schedule / Inactivate Task / Add to Timeline / Reassign To / Select to End / Select All Assignments on This Task	No

	Screen Name	Panes	Split Verti-cally	Details Options	Format Column Options
16	Timeline	Top Only	No	Copy Timeline ▸ Detailed Timeline Zoom to Screen Text Styles... Insert Task ▸ ✓ Show Timeline	No

The following table displays the Screens and provides some background on each:

Screen Name	Note and/or Screen Dumps
1. Calendar	May be displayed in top pane only. 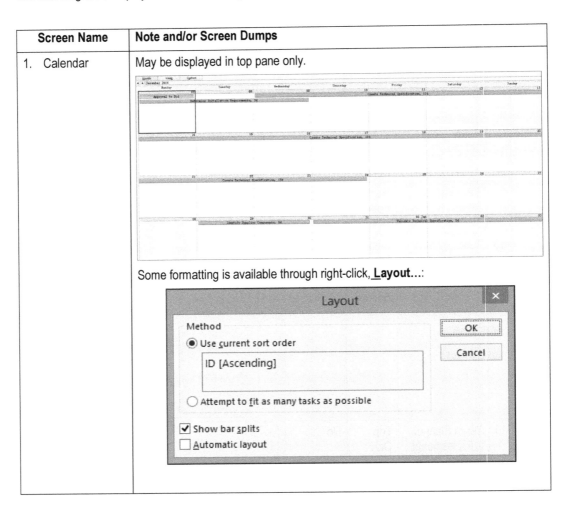 Some formatting is available through right-click, **Layout...**:

Screen Name	Note and/or Screen Dumps
2. Gantt Chart	Designed to be displayed in the top pane: 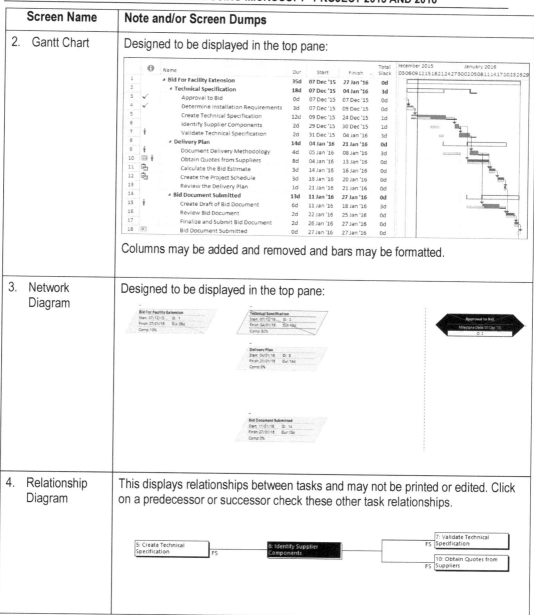 Columns may be added and removed and bars may be formatted.
3. Network Diagram	Designed to be displayed in the top pane:
4. Relationship Diagram	This displays relationships between tasks and may not be printed or edited. Click on a predecessor or successor check these other task relationships.

Screen Name	Note and/or Screen Dumps
5. Resource Form	Designed for the bottom pane: 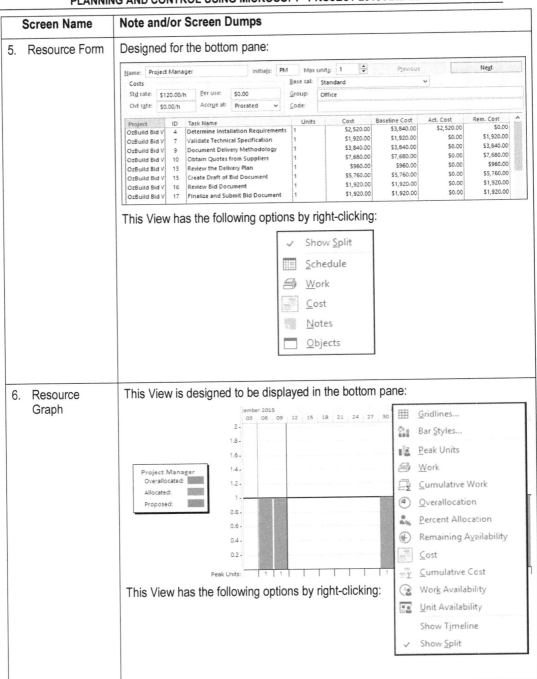
6. Resource Graph	

Screen Name	Note and/or Screen Dumps
7. Resource Name Form	This screen is designed for the bottom pane with the **Resources Sheet** view in the Top pane:

Name:	Project Manager						Previous			Next	
Project	ID	Task Name	Units		Cost	Baseline Cost		Act. Cost		Rem. Cost	
OzBuild Bid V	4	Determine Installation Requirements	1		$2,520.00	$3,840.00		$2,520.00		$0.00	
OzBuild Bid V	7	Validate Technical Specification	1		$1,920.00	$1,920.00		$0.00		$1,920.00	
OzBuild Bid V	9	Document Delivery Methodology	1		$3,840.00	$3,840.00		$0.00		$3,840.00	
OzBuild Bid V	10	Obtain Quotes from Suppliers	1		$7,680.00	$7,680.00		$0.00		$7,680.00	
OzBuild Bid V	13	Review the Delivery Plan	1		$960.00	$960.00		$0.00		$960.00	
OzBuild Bid V	15	Create Draft of Bid Document	1		$5,760.00	$5,760.00		$0.00		$5,760.00	
OzBuild Bid V	16	Review Bid Document	1		$1,920.00	$1,920.00		$0.00		$1,920.00	
OzBuild Bid V	17	Finalize and Submit Bid Document	1		$1,920.00	$1,920.00		$0.00		$1,920.00	

This View has the following options by right-clicking:

✓ Show Split

▦ Schedule

🖨 Work

Cost

Notes

Objects

Screen Name	Note and/or Screen Dumps
8. Resource Sheet	This screen may be used in the top or bottom pane and the columns may be formatted with the **Table and Columns** functions:

	ⓘ	Resource Name	Type	Material Label	Initials	Group	Max. Units	Std. Rate	Base Calendar	Accrue At	Ovt. Rate
1	◈	Project Manager	Work		PM	Office	1	$120.00/hr	Standard	Prorated	$0.00/hr
2		Systems Engineer	Work		SE	Office	1	$90.00/hr	Standard	Prorated	$0.00/hr
3		Project Support	Work		PS	Site	1	$80.00/hr	6-Day Working Week	Prorated	$0.00/hr
4		Purchasing Officer	Work		PO	Office	1	$70.00/hr	Standard	Prorated	$0.00/hr
5		Clerical Support	Work		CS	Office	1	$50.00/hr	Standard	Prorated	$0.00/hr
6		Specialist Consultant	Cost		SC	Contractor				Prorated	
7		Report Binding	Material	Each	RB			$100.00		Prorated	

Screen Name	Note and/or Screen Dumps
9. Resource Usage	This screen may be used in the top or bottom pane: 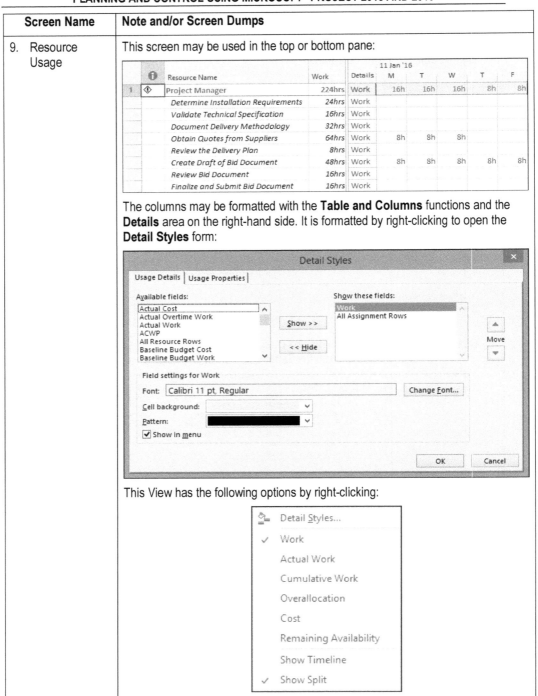

The columns may be formatted with the **Table and Columns** functions and the **Details** area on the right-hand side. It is formatted by right-clicking to open the **Detail Styles** form:

This View has the following options by right-clicking:

This publication is only sold as a bound book, no parts may be reproduced by any means, e.g. electronic, video or print.

© **Eastwood Harris Pty Ltd**   374

Screen Name	Note and/or Screen Dumps
10. Task Details Form	This form is best displayed in the bottom pane and displays information about the task highlighted in the top pane: 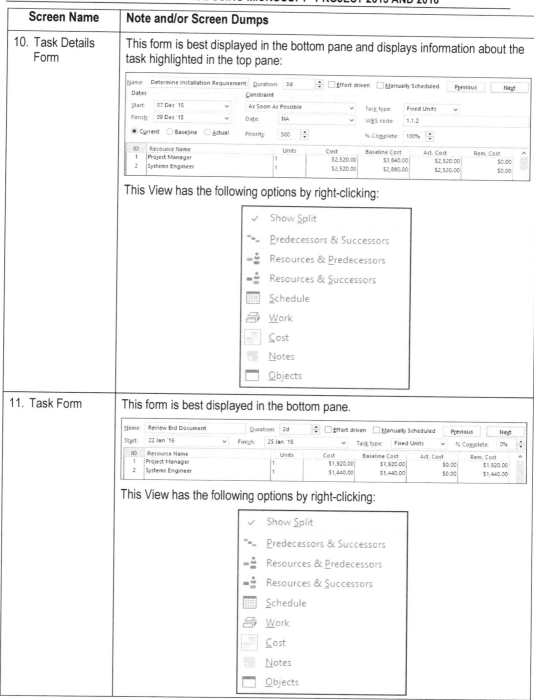 This View has the following options by right-clicking:
11. Task Form	This form is best displayed in the bottom pane. This View has the following options by right-clicking:

Screen Name	Note and/or Screen Dumps
12. Task Name Form	This form is best displayed in the bottom pane.

Name:	Review the Delivery Plan			Previous		Next	
ID	Resource Name	Units	Cost	Baseline Cost	Act. Cost	Rem. Cost	
1	Project Manager	1	$960.00	$960.00	$0.00	$960.00	
2	Systems Engineer	1	$720.00	$720.00	$0.00	$720.00	

This View has the following options by right-clicking:

- ✓ Show Split
- Predecessors & Successors
- Resources & Predecessors
- Resources & Successors
- Schedule
- Work
- Cost
- Notes
- Objects

13. Task Sheet	This View is similar to the Gantt but without displaying the bars. The columns may be formatted using the **Table** and **Columns** functions.

	ⓘ	Name	Dur	Start	Finish	Total Slack	Cost	Work	Predecessc
1		**Bid For Facility Extension**	35d	07 Dec '15	27 Jan '16	0d	$59,260.00	560hrs	
2		**Technical Specification**	18d	07 Dec '15	04 Jan '16	3d	$24,000.00	192hrs	
3	✓	Approval to Bid	0d	07 Dec '15	07 Dec '15	0d	$0.00	0hrs	
4	✓	Determine Installation Requirements	3d	07 Dec '15	09 Dec '15	0d	$5,040.00	48hrs	3
5		Create Technical Specification	12d	09 Dec '15	24 Dec '15	1d	$14,480.00	96hrs	4
6		Identify Supplier Components	2d	29 Dec '15	30 Dec '15	1d	$1,120.00	16hrs	5
7	ⓘ	Validate Technical Specification	2d	31 Dec '15	04 Jan '16	3d	$3,360.00	32hrs	6
8		**Delivery Plan**	14d	04 Jan '16	21 Jan '16	0d	$21,520.00	224hrs	

Screen Name	Note and/or Screen Dumps
14. Task Usage	This view displays the tasks and assignments in the columns and the resources and work on the right-hand side:

	Task Name	Work	Details	M	T	W	T	F
				07 Dec '15				
1	**Bid For Facility Extension**	**560hrs**	Work	16h	16h	24h	8h	8h
2	**Technical Specification**	**192hrs**	Work	16h	16h	24h	8h	8h
3	Approval to Bid	0hrs	Work					
4	Determine Installation Requirements	48hrs	Work	16h	16h	16h		
	Project Manager	*24hrs*	Work	8h	8h	8h		
	Systems Engineer	*24hrs*	Work	8h	8h	8h		
5	Create Technical Specification	96hrs	Work			8h	8h	8h
	Systems Engineer	*96hrs*	Work			8h	8h	8h

The columns may be formatted using the **Table** and **Columns** functions.

This View has the following options by right-clicking:

- Detail Styles...
- ✓ Work
- Actual Work
- Cumulative Work
- Baseline Work
- Cost
- Actual Cost
- Show Timeline
- ✓ Show Split

The formatting of the right-hand side of the screen may be changed by right-clicking to open the **Detail Styles** form:

Detail Styles

Usage Details | Usage Properties

Available fields:
- Actual Cost
- Actual Fixed Cost
- Actual Overtime Work
- Actual Work
- ACWP
- All Task Rows
- Baseline Budget Cost

Show these fields:
- Work
- All Assignment Rows

Show >>   << Hide   Move ▲ ▼

Field settings for Work

Font: Calibri 11 pt, Regular   Change Font...

Cell background:

Pattern:

☑ Show in menu

OK   Cancel

Screen Name	Note and/or Screen Dumps
15. Team Planner	This view displays the resources on the left-hand side and the tasks assignments on the right: 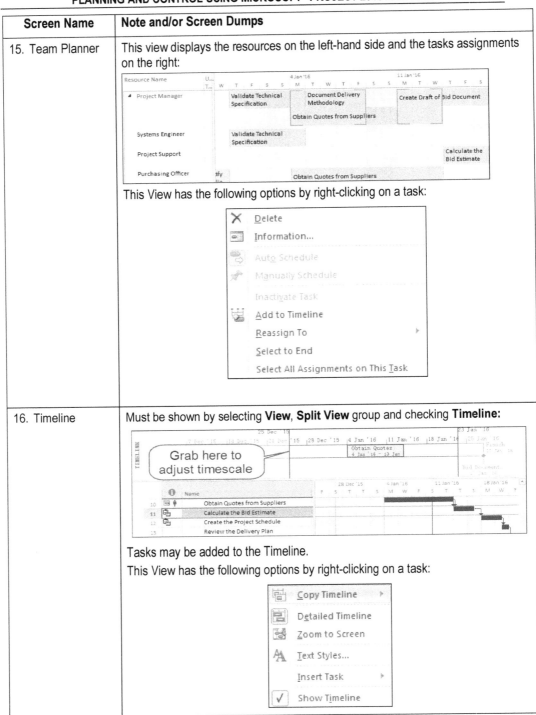 This View has the following options by right-clicking on a task: ✕ Delete 🖼 Information... 🔄 Auto Schedule 📌 Manually Schedule Inactivate Task 📋 Add to Timeline Reassign To ▸ Select to End Select All Assignments on This Task
16. Timeline	Must be shown by selecting **View**, **Split View** group and checking **Timeline:** Grab here to adjust timescale Tasks may be added to the Timeline. This View has the following options by right-clicking on a task: 📋 Copy Timeline ▸ 📋 Detailed Timeline 📋 Zoom to Screen A Text Styles... Insert Task ▸ ✓ Show Timeline

27 INDEX

© **Eastwood Harris Pty Ltd**

ON-LINE TRAINING

INTRODUCTION

Primaskills Pty Ltd is introducing self-paced on-line training courses. The courses are available for both individuals and organizations and can be ordered through www.primaskills.com.au.

These courses are based on the Eastwood Harris materials and are be used in conjunction with the Eastwood Harris books.

COURSE STRUCTURE AND CONTENT

These courses are downloadable recordings of the current Eastwood Harris Project Management Institute accredited training courses and include high quality videos featuring:

- PowerPoint slide presentations covering the concepts and theory behind the software,
- Demonstrations of the software functionality,
- A walk through of the student workshops.

TRAINING ENVIRONMENT

Primaskills can also provide hosted training environment to be used in conjunction with the on-line content.

CORPORATE LICENSES

Corporate licenses available, so all you employees may access Eastwood Harris On-Line training and please contact Primaskills for more information.

AFFILIATE TRAINING COMPANIES

Opportunities also exist for organizations wishing to become a reseller of the Primaskills training content, through an affiliate arrangement.

CONTACT INFORMATION

For additional information on the Primaskills reseller affiliates program, please contact Primaskills on info@primaskills.com.au.

www.primaskills.com.au

Eastwood Harris Pty Ltd, P.O. Box 4032, Doncaster Heights, 3109, Victoria, Australia

AUSTRALIA: Tel: 04 1118 7701   **INTERNATIONAL:** + 61 4 1118 7701

Skype & Email: harrispe@eh.com.au   **Web:** http://www.eh.com.au

EASTWOOD HARRIS

INSTRUCTOR POWERPOINT SHOWS

Eastwood Harris PowerPoint slide shows allow companies and training organizations to run their high quality training are available in PDF and PowerPoint format.

- Most courses are registered with the Project Management Institute, and PMI® Registered Education Providers may award PDUs to attendees.

- The PowerPoint slide presentation is editable, so you may add your logos and tailor the training course to suit your students.

- The PowerPoint slide presentation has instructor notes and white board examples as hidden slides to assist your instructors to understand the material and software.

 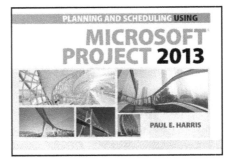

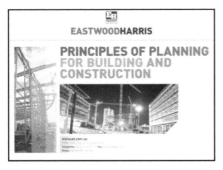 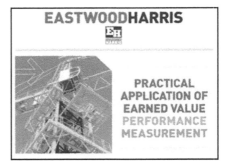

Eastwood Harris Pty Ltd **Web:** www.eh.com.au   **Tel** +61 (0)4 1118 7701   **Email:** harrispe @eh.com.au

Lightning Source UK Ltd.
Milton Keynes UK
UKHW03f0108140518
322543UK00012B/307/P